70th Northeastern Weed Science Society Annual Meeting 2016

1st Annual Northeast Plant, Pest, and Soils Conference

Proceedings – Volume 70

Philadelphia, Pennsylvania, USA
3-7 January 2016

Editor:

Arthur E. Gover

ISBN: 978-1-5108-3023-3

Printed from e-media with permission by:

Curran Associates, Inc.
57 Morehouse Lane
Red Hook, NY 12571

Some format issues inherent in the e-media version may also appear in this print version.

Printed by Curran Associates, Inc. (2016)

Northeastern Weed Science Society (NEWSS)
P.O. Box 25
Woodstown, NJ 08098
USA

Phone: 315-209-7580

northeasternweedscience@hotmail.com

Additional copies of this publication are available from:

Curran Associates, Inc.
57 Morehouse Lane
Red Hook, NY 12571 USA
Phone: 845-758-0400
Fax: 845-758-2633
Email: curran@proceedings.com
Web: www.proceedings.com

NEPPSC

NORTHEASTERN PLANT, PEST AND SOILS CONFERENCE

PROCEEDINGS - VOLUME 1

THE NORTHEASTERN PLANT, PEST, AND SOILS CONFERENCE 2016

A joint meeting of:

Northeastern Weed Science Society

Entomological Society of America Eastern Branch

APS Northeastern Division

American Society for Horticultural Science Northeast Region Founded 1903

American Society of Agronomy | Crop Science Society of America | Soil Science Society of America

Philadelphia, PA; January 3-7, 2016

Proceedings

of the

First Annual Meeting

of the

Northeastern Plant, Pest, and Soils Conference

Arthur E. Gover, Editor

NORTHEASTERN WEED SCIENCE SOCIETY

2016 Sustaining Members

Platinum

BASF The Chemical Company

VALENT

syngenta

Gold

FMC

pbi/gordon corporation
An Employee-Owned Company

DOW Dow AgroSciences

BAYER Bayer CropScience

MONSANTO

Silver

For-Shore Weed Control, Inc.

Gowan USA
The Go To Company

ohp Partners with solutions

DUPONT

UPI

GDM Gylling Data Management

Bronze

Arysta LifeScience
BAAR Scientific, LLC
Crop Management Strategies
ACDS Research
LABServices
TeeJet Technologies
Weeds, Inc.

Entomological Society of America-Eastern Branch

Sponsors

AMVAC

BASF Corporation

Bayer CropScience

Delaware Department of Agriculture

Dow AgroSciences

DuPont Agricultural Products

FMC Corporation

LABServices

Monsanto

Nichino America, Inc.

Suterra, LLC

Syngenta Crop Protection, Inc.

Trece, Inc.

UPI

Wildscape, Inc

PARTICIPATING SOCIETIES

Northeastern Weed Science Society (Host Society)

Entomological Society of America – Eastern Branch

American Phytopathological Society - Northeastern Division

Northeastern Regional Branch of American Society of Agronomy, Crop Science Society of America, & Soil Science Society of America

American Society of Horticultural Science - Northeast Division

SOCIETY PRESIDENTS

Rakesh Chandran, NEWSS President	*rschandran@mail.wvu.edu*
Tracy Leskey, EB-ESA President	*tracy.leskey@ars.usda.gov*
Michael Fidanza, NE-ASA/CSSA/SSSA President	*fidanza@psu.edu*
Francis Ferrandino, NED-APS President	*francis.ferrandino@ct.gov*
Elsa Sanchez, NE-ASHS President	*ess11@ag.psu.edu*

NEPPSC PROGRAM COMMITTEE

Shawn Askew, Program Chair & NEWSS Rep.	*saskew@vt.edu*
Anne Nielsen, Program Co-Chair & EB-ESA Rep.	*nielsen@aesop.rutgers.edu*
Michael Fidanza, NE-ASA/CSSA/SSSA Rep.	*fidanza@psu.edu*
Margaret T. McGrath, NED-APS Program Rep.	*mtm3@cornell.edu*
Elsa Sanchez, NE-ASHS Program Rep.	*ess11@ag.psu.edu*

Table of Contents

URBAN AND MEDICAL PESTS SYMPOSIUM

BIOLOGICIAL CONTROL OF WEEDS SYMPOSIUM

INSECT DETECTION, EVALUATION, AND PREDICTION SYMPOSIUM

RESEARCH POSTERS - STUDENT CONTEST

NEWSS STUDENT CONTEST - ORAL PRESENTATIONS (I)

EB-ESA PHD STUDENT CONTEST - ORAL PRESENTATIONS

NE-ASA/CSSA/SSSA STUDENT CONTEST - ORAL PRESENTATIONS

NED-APS STUDENT CONTEST - ORAL PRESENTATIONS

EB-ESA BS/MS STUDENT CONTEST - ORAL PRESENTATIONS

RESEARCH POSTERS

ORGANIC PRODUCTION, BENEFICIAL ORGANISMS, AND BIOLOGICAL CONTROL

AGRONOMIC CROPS (I)

TURF AND ORNAMENTALS

VEGETATION MANAGEMENT AND RESTORATION

FRUIT CROPS

SOIL AND ENVIRONMENTAL QUALITY

INVASIVE SPECIES

VEGETABLE AND HERB CROPS (I)

AGRONOMIC CROPS (II)

TURF AND ORNAMENTALS

VEGETABLE AND HERB CROPS (II)

URBAN AND MEDICAL PESTS SYMPOSIUM

BED BUGS AS VECTORS OF *TRYPANOSOMA CRUZI*, THE ETIOLOGIC AGENT OF CHAGAS DISEASE. M.Z. Levy*, R. Salazar, R. Castillo-Neyra, A. Tustin, K. Borrini-Mayori, and C. Naquira, University of Pennsylvania, Philadelphia, PA (3)

ABSTRACT

Through a series of experiments we show that bed bugs (*Cimex lectularius*) can acquire *Trypanosoma cruzi* by feeding on infected mice, transmit it back to susceptible animals during cohabitation (in which the mice ate them) and by the usual route (through their feces). Whether bed bugs are, or will become, epidemiologically important vectors of the parasite remains unclear. To address the issue we develop a modified Ross-MacDonald model of vector-borne *T. cruzi* transmission and applied it to both a traditional vector, *Triatoma infestans*, and *Cimex lectularius*. In relation to some key parameters of the model, such as the ratio of vectors to hosts and the biting rate of the vector, bed bugs are more worrying than *T. infestans*. We consider the death rate of the vector, and run experiments to assess whether *T. cruzi* might increase mortality among bed bugs or triatomines. It does not. We consider the range of hosts and feeding preferences of each insect. We show that triatomines can only support sustained *T. cruzi* transmission when exposed to a rather complex host community; we discuss whether such conditions exist in houses, hotels, and other environments affected by bed bugs in the US and abroad.

HUMAN NETWORKS (CITIES AND HIGHWAYS) DRIVE THE RAPID EVOLUTION OF DISEASE VECTORS. D.M. Fonseca*, A. Egizi, and N. Fefferman, Rutgers University, New Brunswick, NJ (4)

ABSTRACT

Invasive mosquitoes have caused inordinate human suffering because they are very abundant in their exotic ranges and can drive local or exotic disease epidemics. Since invasiveness often correlates with high genetic diversity it was expected that invasive mosquitoes would be genetically very diverse, but recent comparisons of native and exotic populations using high-resolution molecular markers revealed they are not. Invasive mosquitoes often exploit domestic environments, and the resulting extensive overlap with humans significantly enhances their effectiveness as vectors of human diseases. It is unclear, however, whether domestication was a pre-condition for invasiveness or instead a consequence of it. To understand factors that deter or promote invasiveness we studied a new invasive mosquito in North America and Europe and in its native East Asia. We traced the source of all exotic populations to central Japan, where we found high genetic diversity but homogeneity across populations. Spatiotemporal analyses of exotic populations revealed that (1) genetic diversity declined precipitously outward from introduction points limiting expansion; (2)

populations across the exotic range had very different genetic signatures; (3) all broadly expanding populations (=invasive) had genetic signatures resulting from admixture of separate introductions. We propose that post-introduction human-assisted rapid movement of specimens across the exotic range (rescued) self-limiting introductions by mixing locally differentiated genotypes. Therefore invasiveness is not a required pre-existing trait in invasive mosquitoes, but instead can evolve in the exotic range leading inexorably to dangerous human disease vectors.

HARBORING SECRETS: PATHOGENS AND ECTOPARASITES OF NEW YORK CITY RATS. M.J. Frye*, Cornell University, Elmsford, NY (5)

ABSTRACT

The Norway rat (*Rattus norvegicus*) is a commensal rodent that lives in close proximity to humans in urban environments, is a reservoir of zoonotic pathogens and host to ectoparasitic arthropods. Here, we report on the diversity and abundance of ectoparasitic arthropod species and associated pathogenic bacteria from 133 Norway rats trapped over a 10-mo period in Manhattan, NY. Norway rats were host to the tropical rat mite [*Ornithonyssus bacoti* (Hirst)], the spiny rat mite (*Laelaps echidnina* Berlese), *Laelaps nuttalli* Hirst, the spined rat louse [*Polyplax spinulosa* (Burmeister)], and the Oriental rat flea [(*Xenopsylla cheopis*) (Rothschild)]. A flea index of 4.1 *X. cheopis* was determined, whereas previous studies in New York City reported 0.22 fleas per rat. Multiple species of pathogenic *Bartonella* were identified from Oriental rat fleas that were related to *Bartonella tribocorum*, *Bartonella rochalimae*, and *Bartonella elizabethae*. However, no evidence of *Yersinia pestis* or *Rickettsia* spp. infection was detected in fleas.

ROLE OF *DEMODEX FOLLICULORUM* AND *DEMODEX BREVIS* IN PARTHENOGENESIS OF BLEPHRITIS IN NORTH INDIA. N.K. Gill*, Punjabi University, Patiala, India (8)

ABSTRACT

The aim of the present study was to evaluate the role of *Demodex folliculorum* and *Demodex brevis* in parthenogenesis of Blepharitis in North India. Total 300 patients between the age group of 30-70 years from the Eye department of Sri Guru Ram Das Medical College, Amritsar (Punjab) and Post Graduate Institute of Medical Science and Research Chandigarh (Covering the states of Haryana, Punjab, Jammu and Kashmir, and Uttarakhand) were included in the study. All the patients were suffering from itching, redness, watery eyes, foreign body sensation, heavy eyelids, presence of cylindrical dandruff and blurry vision. 55 percent of the patients were found to be infested with *Demodex* spp. The association between the presence of *Demodex* spp. and host factors were analyzed on the basis of self-made questionnaire (age, sex, demodicosis symptoms). The samples were considered positive if at least one parasite was found in the eyelashes. It has been observed that *Demodex folliculorum* infection was more frequently found than the *Demodex brevis.* The highest percentage of the infection was reported between the age group of 50-70 years belonging to low socioeconomic status.

BIOLOGICIAL CONTROL OF WEEDS SYMPOSIUM

WHAT IS REALLY A RISK AND WHAT IS NOT? T. McAvoy* and S. Salom, Virginia Tech, Blacksburg, VA (12)

ABSTRACT

A review of non-target impacts from deliberately introduced biological control agents of weeds found non-target impacts to be rare (Suckling and Sforza, 2014). The impact of 43 species of biological control agents on 140 non-target plant species had no significant impact on non-target plant populations. This is a safety record of over 99 percent. Only two agents, *Cactoblastis cactorum* and *Rhinocyllus conicus* have had large impacts on native cacti and thistles, respectively. Based on the current criteria for releasing biological control agents these two species would not be released today. Due to improved science, concern for native species especially threatened, and endangered species and consideration of ecosystem effects, current and future negative non-target impacts are greatly reduced. However, in the last few years the approval rate for releasing biological control agents has been low. Possibly due to an overcautious concern of negative non-target effects and lack of benefit analysis. Several biological control agents, that were released in the past and have significantly contributed to the management of major weeds with little to no negative impacts, would have been rejected for release using current standards for release approval (Hinz et al. 2014). More robust documentation and publicity of the benefits and explanation of negative impacts if any, would improve scientific and public acceptance of classical biological control.

DEMOGRAPHIC MODELING IN WEED BIOCONTROL. L. Milbrath*, A.S. Davis, and J. Biazzo, USDA-ARS, Ithaca, NY (13)

ABSTRACT

Demographic matrix modeling of plant populations can be a powerful tool to identify key life stage transitions that contribute the most to population growth of an invasive plant and hence should be targeted for disruption. Therefore, this approach has the potential to guide the pre-release selection of effective biological control agents. We have parameterized a five life-stage matrix model for the swallow-wort biological control program. Pale swallow-wort, *Vincetoxicum rossicum*, and black swallow-wort, *Vincetoxicum nigrum* (Apocynaceae: Asclepiadoideae), are herbaceous, perennial, milkweed vines introduced from Europe. Both species are invading a variety of natural and managed habitats in eastern North America. Demographic transitions were measured in field studies for both swallow-wort species in open field and, for pale swallow-wort, forest habitats in New York State (N = six populations). Vital rates that were estimated included germination, survival, maturation (growth to the next life stage), and fecundity (viable seeds produced per plant). Elasticity analyses have identified several potentially important transitions for one or both species of swallow-wort: survival of vegetative juvenile, small flowering and large flowering plants; growth of seedlings, juveniles and small flowering plants to the next life stage; and reproduction of small and large flowering plants. In general, a combination of moderate to substantial reductions in survival, fecundity, and/or growth is needed for most populations to decrease in size. Incorporation of available impact data of candidate biological control agents show that control may be promising for forest infestations of swallow-wort but not for most field infestations.

WEED BIOCONTROL IN LANDSCAPE RESTORATION. E.C. Lake* and J.A. Hough-Goldstein, USDA-ARS, Fort Lauderdale, FL (14)

ABSTRACT

Weed biological control programs in natural areas are often undertaken with the goal of restoring native plant communities and/or ecosystem services to a pre-invasion level. These objectives may be achieved in some areas with biological control alone; however, in other sites integration of biological control with additional management techniques is necessary. Biological control has been successfully integrated with herbicides, mechanical control, fire, and grazing to manage invasive weeds. The control technique(s) implemented in a given system may depend on restoration goals, treatment costs, the specific habitats invaded by the weed, and the type of damage caused by the biological control agent. Mile-a-minute weed, (*Persicaria perfoliata* (L.) H. Gross), is an aggressive annual vine that has invaded the eastern United States. In some cases, successful biological control of mile-a-minute weed by the weevil *Rhinoncomimus latipes* Korotyaev has resulted in replacement by other undesirable vegetation, the invasive treadmill effect. Two small-scale experiments in this system successfully integrated biological control with native plantings to restore plant communities. In one experiment, weevil releases were integrated with a single application of a pre-emergent herbicide and plantings of a native forb and tree. Two years later, the cover of native plants in the integrated treatment plots was greater than 80 percent. In a second experiment, the combination of biological control and plantings of a native seed mix resulted in increased richness and diversity of native plants after three years. In both experiments, the integration of techniques reduced the abundance of the target weed, promoted recruitment of additional native plant species that were not included in the plantings, and prevented dominance by other invasive plants compared to non-planted control plots.

PATHOGENS IN WEED BIOCONTROL: OPPORTUNITIES AND LIMITATIONS WITH THE USE OF BIOPESTICIDES. J.C. Neal*, North Carolina State University, Raleigh, NC (17)

ABSTRACT

Plant pathogens have been evaluated for both "classical" and "innundative" biocontrol strategies. Several rust fungi have been introduced as "classical" biocontrol agents with limited success in North America. Numerous pathogens have been evaluated for innundative or "bioherbicidal" applications. Despite significant research in the field, there are very few examples of successful commercializations of pathogens as bioherbicide agents. Constraints to development and commercialization of plant pathogens as biological control agents for weeds have been categorized as biological, environmental, technological, and commercial factors. However, recent product development initiatives have resulted in successful registrations. Because important constraints to commercialization include a need for modified environment conducive to infection and high costs of production, several research groups have emphasized development of biocontrol agents for turfgrass, a high-value site for which irrigation is often available. Two products, containing *Sclerotinia minor* or *Phoma macrostoma,* recently received registrations in Canada for broadleaf weed control in turfgrass systems. At present, neither is commercially available. Recent registration of tobacco mild green mosaic tobamovirus (Solvinix LC) for biological control of tropical soda apple demonstrates that biocontrol with plant pathogens can be effective in agricultural systems with lower cost-points, such as pasture and non-crop areas. One technique which has shown promise is the use of microbes to biomanufacture natural products that are toxic to weedy plants. The utilization of naturally occurring toxins may circumvent several of the constraints commercialization described above.

INSECT DETECTION, EVALUATION, AND PREDICTION SYMPOSIUM

EMERALD ASH BORER IN CONNECTICUT: DETECTION AND MANAGEMENT. C.E. Rutledge*, The Connecticut Agricultural Experiment Station, New Haven, CT (18)

ABSTRACT

Emerald ash borer was first detected in Connecticut in 2012. Since that time it has spread from an initial 5 towns to over 70 towns, and is now found in 7 of Connecticut's 8 counties. Towns in the initial detection area are now experiencing large ash diebacks and demand for expertise in treating and managing emerald ash borer is growing. This talk will review the history of emerald ash borer in the state, the detection modalities used to track its spread, and discuss the disparate response of towns to the problem.

SOUTHERN PINE BEETLE IN THE NORTHEAST - DETECTION AND MANAGEMENT OF A SOUTHERN PEST AT ITS NEW NORTHERN DISTRIBUTION. R.M. Cole* and J.A. Cancelliere, NYS Department of Environmental Conservation, Albany, NY (19)

ABSTRACT

The southern pine beetle *Dendroctonus frontalis* was detected on Long Island in September of 2014. Several months later SPB was also found established at multiple sites in Connecticut. Trapping during the following summer (2015) resulted in detections in upstate New York, Massachusetts and Rhode Island, but no infested trees have been found in those locations.

Aerial and ground surveys on Long Island have revealed that SPB is widely established in Suffolk County, particularly throughout the Long Island Pine Barrens Preserve, where many stands of overstocked and stressed pitch pines exist. These stands vary from pure pitch-pine to mixed oak-pitch pine and occur over multiple landownerships including federal, state, municipal and private lands. SPB has been found predominantly in pitch pine, but also in red, Scots and white pine and Norway spruce. Native hard pines, such as pitch and red, are not dominant in the Northeast, but they often represent unique ecosystems and occupy distinctive niches.

Large scale tree mortality and SPB spot expansion has been documented on Long Island. NYSDEC Forest Health, in cooperation with the Central Pine Barrens Commission and Suffolk County, began suppression efforts in October 2015. Methods developed in southern forests, like cut-and-leave and cut-and-remove suppression, have been considered for use, but due to logistical constraints and the lack of a forest products industry, only the cut-and-leave treatment has been applied to date. Infested trees are being cut and left on the ground early in winter in hopes that the brood will be negatively affected by the tree felling and exposure to cold temperatures. Preventative management, in which forests are thinned to a lower basal area to make them "beetle-proof", is an effective strategy, but the lack of a timber market ensures that costs per acre for felling/removing trees remain too high. DEC is currently attempting a timber sale of high-risk stands within Rocky Point State Forest, to test whether a market may exist and if this practice can be utilized in the future.

Many research questions are being addressed to learn how SPB is behaving in the north. We are looking at patterns of spot expansion and tree fade, the relationship between basal area and stand susceptibility to SPB, host preferences, and the role of cold winter temperatures on SPB mortality and subsequent rate of spread. The infestation on Long Island will provide essential baseline information for land managers throughout the Northeast preparing to deal with this devastating insect.

WINTER MOTH IN THE NORTHEAST AND ITS ASSOCIATED PARASITOIDS, PATHOGENS, AND PREDATORS. H.J. Broadley*, G. Boettner, and J. Elkinton, University of Massachusetts, Amherst, MA (21)

ABSTRACT

Winter moth (*Operophtera brumata*), a polyphagous geometrid, was accidently introduced to the northeastern United States from Europe in the 1990s. Since its introduction, winter moth has exhibited population outbreaks reaching densities of up to 500 pupae/m^2 and resulting in 50 percent defoliation. It has spread coastal at about 8.3 km/year throughout eastern Massachusetts, north into Maine, and south onto the Cape, across Rhode Island and most recently to Connecticut. To control winter moth, the Elkinton lab began releases of the tachinid parasitoid *Cyzenis albicans* beginning in 2005. *C. albicans* has now been released at 40 sites and has established at 17 sites. In Welleseley, MA parasitism has reached almost 60 percent at one site and has spread over an area of 9 km^2 resulting in a notable decline of winter moth densities. We are testing for a synergistic relationship with parasitism by *C. albicans* and with the native pupal predator community. We have found limited evidence of a synergism of *C. albicans* parasitism and pupal predators, but we have found high parasitism rates (up to 33 percent) by a native ichenumonid wasp, *Pimpla aequalis*. We suggest that the aggregation of this wasp after winter moth establishment may be responsible for the apparent synergistic effect. Additionally, to assess other potential sources of native natural enemies, we are evaluating regulatory agents acting on winter moth's native congener Bruce spanworm (*O. bruceata*), which rarely outbreaks. We have found that the fungal pathogen microsporidia appears to be an important regulator of Bruce spanworm, but this time is not abundant in the winter moth population. Nucleopolyhedrovirus (NPV) does not appear to inflict significant mortality in either moth species. Additionally, we have found tachinid and ichneumonid parasitoids in Bruce spanworm and are testing their phylogenetic relationship to winter moth's parasitoids. To understand winter moth's population dynamics, it is essential that we understand the complex dynamic of native and non-native parasitoids, pathogens, and predators acting upon winter moth and its sympatric sister species Bruce spanworm.

THE VELVET LONGHORNED BEETLE (*TRICHOFERUS CAMPESTRIS*): PEST STATUS, INTERCEPTIONS, AND ITS POTENTIAL IMPACT ON EASTERN FORESTS. J.D. Young*, USDA APHIS PPQ, Baltimore, MD (22)

ABSTRACT

The Velvet Longhorned beetle (*Trichoferus campestris*) is an exotic cerambycid that is native to Asia. The species has been accidently introduced to just a few locations in the US. The biology of this pest, trapping and survey options and interceptions in international conveyance will be discussed.

RESEARCH POSTERS - STUDENT CONTEST

WATER RELATIONS OF TREES GROWING IN A GREEN INFRASTRUCTURE (GI) STORM WATER TRENCH. R. Galanti*, S. Olshevski, and S. Eisenman, Temple University, Ambler, PA (25)

ABSTRACT

Storm water management is a major concern for cities with outdated combined sewer systems. Uncontrolled storm water pollutes urban watersheds, and impairs the ecological function of streams and rivers. Storm water interception using urban green infrastructure (GI) installations planted with street trees will be a major contributor to improving overall runoff control. Evapotranspiration by trees should contribute greatly to diverting excess water from sewer systems. It is important to assess the performance of these trees to better understand their contributions to storm water management.

This research project focuses on water relations of trees in a GI tree trench system in the Mt. Airy section of Philadelphia. *Acer rubrum* 'Armstrong' and *Platanus* × *acerifolia* 'Bloodgood' were evaluated from May through November for stomatal conductance, leaf water potential (Ψ_{lf}) and leaf area index (LAI). Water relation trends were evident, and a one-way ANOVA with post-hoc Tukey HSD test showed a significant difference between stomatal conductance rates of the two species, but not within species. Analysis of Ψ_{lf} data was performed using a Kruskal-Wallis test rank sum test followed by Dunn's test of multiple comparisons. These analyses also showed significant differences between the ranked data of the two species. In general, through the entire growing season, *P.* × *acerifolia* had greater stomatal conductance and lower susceptibility to water stress than *A. rubrum* 'Armstrong'. This study provides some fundamental data on water relations and species performance in a GI tree trench system.

COMPARING STOMATAL CONDUCTANCE OF STREET TREE SPECIES IN TRADITIONAL TREE PITS AND GREEN INFRASTRUCTURE STORM WATER TRENCHES. S. Olshevski*, R. Galanti, and S. Eisenman, Temple University, Philadelphia, PA (26)

ABSTRACT

Urban combined sewer systems (CSS) such as that of Philadelphia can lead to combined sewer overflows (CSO) during intense storm events. CSO generate serious environmental concern due to their introduction of untreated waste into local waters. Philadelphia's Green City, Clean Waters initiative calls for significant implementations of green infrastructure in order to control the amount of stormwater runoff entering the sewer system. One such structure is a stormwater tree trench, which consists of soil-filled tree pits installed on top of a large gravel-filled catchment area beneath the sidewalk. Stormwater, collected through curbside inlets, is distributed through the catchment area via a perforated pipe. The trenches are designed to retain large volumes of water, with a portion eventually percolating into the soil below, and some being accessed and transpired by the trees above. In order to assess the effectiveness of the stormwater trenches in controlling CSO, it is necessary to evaluate the water relations of the various associated tree species. Stomatal conductance measures the concentration of water vapor released from tree leaves, and thus it serves as an excellent proxy for total evapotranspiration as well as water stress.

The stomatal conductance of 25 trees of 13 different species/cultivars located on two blocks in the Mt. Airy section of Philadelphia was monitored from June-October 2015. A Kruskal-Wallis test showed that the mean ranks of conductance data by species were significantly different. A Dunn's test of multiple comparisons using rank sums was conducted to determine which groups/species differed significantly. The results suggest that some species may be better suited for use in urban GI trench trees than others. Two potential benefits of these species are 1) they have higher overall levels of conductance resulting in greater water movement out of the system via evapotranspiration following rain events, and 2) they may be less susceptible to water stress during periods of low precipitation. Furthermore, pairwise t-tests revealed that *Koelreuteria paniculata* and *Prunus sargentii* trees conducted significantly more water within stormwater trenches compared to traditional, isolated tree pits, whereas *Quercus macrocarpa* trees showed the opposite tendency. This research provides useful insight for further development of stormwater tree trenches by showing that tree species selection can influence the effectiveness of the system.

HOPS (*HUMULUS* SPP.): A POTENTIAL VALUE-ADDED CROP FOR NJ FARMERS.
R. Pyne*, M. Muehlbauer, E. Dager, W. Cowgill, J. Grande, D. Giurleo, and J. Simon, Rutgers University, New Brunswick, NJ (28)

ABSTRACT

Hops (*Humulus lupulus*) are herbaceous, perennial vines cultivated for their female flowers or cones, which are a critical ingredient in beer-making. A burgeoning craft brewery industry has been the impetus for the rise in demand for hops. Although Mid-Atlantic US states account for an estimated 34 percent of national brewery sales, virtually no hops are grown in this region. As interest in craft brewing and sourcing local ingredients rise, a unique opportunity has emerged for Mid-Atlantic farmers to supply this high value crop. As farmers develop an interest in growing hops, it is important to understand the inputs needed to establish and grow this niche crop in accordance with industry standards. A SARE Partnership Grant-supported study was initiated at Rutgers University to establish a centrally located demonstration hopyard at the Clifford E & Melda C. Synder Research Center for Sustainable Agriculture, Pittstown, NJ to evaluate the performance of 10 commonly grown hop varieties. In addition, hop samples were donated from hop growers throughout New Jersey. Analysis of alpha/beta acid content by American Society for Brewing Chemists (ASBC) protocol 'Hops-14' and essential oil composition by 'Hops-17' provided the essential chemical metrics. Interestingly, major essential oil components (β-myrcene, α-humulene and β-caryophyllene) varied within the same variety at different locations, which provides interesting implications through potential new flavor profiles for local brewers.

WHERE TO SAMPLE? AN ANALYSIS OF WHOLE FARM CORN STALK NITRATE TEST (CSNT) RESULT VARIABILITY AND HOW FARMERS CHOOSE FIELDS TO SAMPLE. R.S. Breslauer*, P. Berenguer, and Q.M. Ketterings, Cornell University, Ithaca, NY (30)

ABSTRACT

Nitrogen (N) is the most commonly limiting nutrient in corn (*Zea mays* L.) production in many agricultural areas. Dairy farmers that grow corn for silage or grain are very interested in applying adequate amounts of N to fields to reach yield potentials. The Corn stalk Nitrate Test (CSNT) is a post-season plant evaluation of N-supply to the crop relative to crop needs. This test can be used to aid in an adaptive N management program for corn growers that aims to improve N use efficiency over time. Results for CSNT are greatly impacted by soil N availability which can be variable from field to field and from farm to farm. The objectives of this study are to (1) analyze whole farm profiles of CSNT's to characterize whole-farm and field-field CSNT variability and (2) evaluate the decision-making process of farmers in choosing which fields to sample for CSNT, realizing that stalk sampling of all corn fields on a farm is typically not feasible. Corn stalk samples were collected on eight NY farms between 2009 and 2011. Farms were located in Cayuga, Clinton, St. Lawrence, Lewis, Jefferson, and Franklin counties. Stalks were sampled from 15 to 35 cm (6 to 14 in) above the ground at a minimum sampling density of 2.5 stalks per hectare (1 stalk per acre). Two of the farms were interviewed to identify criteria for selection of fields that would best represent their farm if given a limitation on the number of fields that could be sampled (2, 4, 6 fields, and 50 percent of their fields versus all fields). Variability in distribution of acreage and number of fields in each CSNT interpretation class was large. Percent of acres in excess on a farm ranged from 37-79 percent suggesting that whole-farm CSNT distribution is indeed farm specific. Despite this variability in distribution, both farms that were interviewed showed a desire to select fields with on opposite sites of the N supply spectrum. Partial sampling of farm G was not representative of the whole farm distribution whereas partial sampling of farm H was representative of the whole farm distribution. This is because farm G was skewed towards excess values whereas farm H had a more uniform CSNT distribution. Overall, accounting for field histories and whole farm distribution of CSNT values is needed to obtain a representative CSNT subsample on a farm.

DETERMINATION OF OPTIMUM NITROGEN RATES FOR WINTER CEREALS USING YIELD RESPONSE CURVES. Q.M. Ketterings, S.E. Lyons, and Z. Tang*, Cornell University, Ithaca, NY (31)

ABSTRACT

Winter cereals such as cereal rye (*Secale cereal L.*), and triticale (*x Triticosecale Wittm.*) can be grown as forage double crops after corn silage in the Northeastern US. This practice has become increasingly popular in recent years, because of soil and nitrogen conservation benefits, reduced crop production risks (two crops instead of one), and the potential to increase overall productivity and economic profits. For optimum economic profit from the production of these winter forages, nitrogen needs to be applied at the most economic rate of nitrogen fertilizer (MERN). Determination of a MERN requires field trials with a minimum of five nitrogen rates. Such trials were conducted in New York in 2013-2015. The yield response data from these trials can be fitted by different statistical models, which typically include linear-plateau, quadratic, quadratic-plateau, exponential and square root models. Data from multiple sites were analyzed to determine the impact of model selection on MERN and predicted yield at MERN. Statistical criteria were used to evaluate the goodness of fit of models. In addition, models were evaluated based on differences in nitrogen use efficiency (NUE) and apparent nitrogen recovery (ANR), as measures of the potential environmental impact associated with the models. Three different price ratios of forage to nitrogen fertilizer were used to evaluate the stability of profits and environmental impacts of different MERNs. Results will be presented for multiple sites with various nitrogen response characteristics.

COVER CROP MIXTURES FOR SUSTAINABLE NO-TILL SWEET CORN PRODUCTION.

J.S. Fine* and M. Hashemi, UMass Amherst, Amherst, MA (32)

ABSTRACT

Cover crop mixtures are increasingly promoted for their diverse benefits. While fall-planted forage radish (*Raphanus sativus* L. var. *longipinnatus*) cover crops have shown successful weed suppression and nutrient cycling, research is lacking on effects of forage radish as a member of a multi-species winter cover crop mixture. This experiment evaluates nutrient cycling, feasibility of early planting and sweet corn yield in a no-till system following cover crop mixtures that include forage radish. In August 2014, three different cover crops were seeded: monoculture forage radish (FR), oats/forage radish (OFR), and peas/oats/forage radish (POFR) and no cover crop (No CC). All cover crops winter-killed. Early maturity sweet corn was no-till planted in May 2015. Three nitrogen fertility treatments examined the synchrony between nutrient release from decomposing cover crops and the uptake by sweet corn. Analysis showed that fall cover crop biomass and plant carbon-to-nitrogen ratio did not differ significantly between cover crop treatments. Cover crop mixtures scavenged fall soil nitrogen effectively compared with no cover crop, preventing winter nitrate leaching. Sweet corn did not respond to additional nitrogen fertilization. Sweet corn yield following all three cover crop treatments were statistically greater than yield following no cover crop. First-year results suggest that OFR could provide optimal synchrony between nitrogen release and sweet corn demand while reducing spring nitrate leaching.

DECOMPOSITION RATE AND RELEASE OF NITROGEN FROM RYE, FORAGE RADISH, OR WINTER PEAS COVER CROPS UNDER CONVENTIONAL OR NO-TILLING SYSTEMS. E. Jahanzad*, A.V. Barker, M. Hashemi, and A. Sadeghpour, University of Massachusetts, Amherst, MA (33)

ABSTRACT

Increased frequency of continuous cropping and extensive use of synthetic fertilizers has threatened environment and human health and has increased costs of crop production. Employing nitrogen management practices and use of cover crops could be a possible alternative that may contribute to more sustainable environment and farming systems via more efficient nutrient cycling and less fertilization. To study the efficiency of three different cover crops from different plant families under no- till or conventional tillage conditions, a 2-yr field experiment was conducted. Cover crops used in this study included annual winter rye, forage radish, and Austrian winter peas. Our results indicated that forage radish produced higher dry matter compared with rye and winter peas. All three cover crops species had a faster decomposition rate and release of nitrogen under conventional tilling system. During the decomposition process, forage radish and peas lost their initial biomass and nitrogen content at a faster trend than winter rye. Considering the decomposition rate and trend of nitrogen release in cover crops, forage radish and peas may be a more feasible option than rye to provide sufficient N for the early planted cash crops in the spring.

ANTIFEEDANT EFFECTS OF MINERAL OIL ON COLORADO POTATO BEETLE. A.K. Galimberti* and A. Alyokhin, University of Maine, Orono, ME (34)

ABSTRACT

Mineral oil can be an important part of IPM programs due to its efficacy against a variety of pests and its reduced impact on natural enemies. In addition to its insecticidal properties, mineral oil also has demonstrated repellant effects against some insects, deterring feeding and oviposition. Mineral oil is used in potato to control potato virus Y, but its effect on Colorado potato beetle has not been studied. In this experiment, we tested the antifeedant activity of mineral oil on Colorado potato beetle using choice and no-choice bioassays. First instar, fourth instar, and adult beetles were allowed to feed on leaves dipped in mineral oil or water. Feeding and location of the beetle was measured after 24 h. In the choice experiment, adults fed significantly more on water-treated leaves than oil-treated leaves. First instars were significantly more likely to be found on water-treated leaves. Fourth instars were not affected. In the no-choice bioassay, first instars were also found significantly more on water-treated leaves, although there was no significant difference in feeding in any stage. The results show that mineral oil may deter feeding by Colorado potato beetle, although beetles will still feed on leaves treated with oil. Information on other ways oil affects potato beetle will help show whether mineral oil can contribute to Colorado potato beetle control as part of an IPM program.

EVALUATING A TRI-PARTITE IPM PROGRAM FOR ONION THRIPS IN ONION. A.B. Leach*, S. Reiners, and B.A. Nault, Cornell University, Geneva, NY (35)

ABSTRACT

Onion thrips (*Thrips tabaci*) is a significant pest of onion worldwide. Growers rely on insecticides to control onion thrips, but other tactics are needed to improve profitability and sustainability of onion production. Promising non-chemical management tactics include reducing nitrogen applied at planting and selecting cultivars with partial thrips resistance. In NY in 2015, combinations of varying levels of nitrogen and insecticide use were evaluated for thrips control in onion varieties that ranged from high to low thrips susceptibility, cvs. 'Bradley', 'Delgado' and 'Avalon'. Within each cultivar, a 3 (nitrogen rate) x 3 (insecticide regimes) factorial was arranged in a RCBD with each factor replicated 5 times. Levels of nitrogen applied at planting were 60, 90, and 125 lbs/acre, whereas insecticide regimes included a weekly spray program (Standard), a spray program in which sprays were based on an action threshold of 1 larva per leaf (IPM), and an untreated control. Thrips densities were recorded weekly and bulbs were harvested in the fall. Regardless of cultivar, nitrogen had no effect on either total thrips densities or onion bulb yields. Across cultivars, insecticide regime had the greatest impact on thrips densities as Standard and IPM spray programs provided equivalent and superior control relative to the untreated check. For 'Bradley' and 'Delgado', bulb yields in Standard and IPM spray programs were similar and averaged 17 grams per bulb more than those in untreated controls; no yield differences occurred for 'Avalon'. Results indicated that growers can reduce fertilizer and insecticide use while maintaining marketable yields.

MONITORING HONEY BEE EXPOSURE TO PESTICIDES ACROSS A GRADIENT OF URBANIZATION. A.F. Thomas*, LIU Brooklyn, Ridgewood, NY (36)

ABSTRACT

North American beekeepers have reported alarmingly high rates of colony loss over the last decade. Multiple factors are thought to play a role in colony loss, of which pesticides have emerged as a particular concern. Analysis of hive components from North American apiaries has revealed that honey bees come into contact with numerous pesticides. However, few studies have assessed honey bee exposure to pesticides in urban settings despite the growth of urban beekeeping and urban agriculture. This study sought to document honey bee interactions with pesticides across an urban-suburban gradient in the New York City metropolitan area. Trapped pollen was collected from beekeepers in Manhattan, Brooklyn and Queens, and suburban locations in Nassau and Westchester counties. LC/MS-MS and GC/MS were used to test pollen for pesticides utilizing a modified QuEChERS technique. Hazard quotients (HQ) were calculated for the combined residues at each field site. Seven total pesticides were found in our study, with an average of one per sample. One site had an HQ value that represented a potential hazard to bees. Comparison of pesticide profiles among our field sampling locations – and between our study and other published studies – suggests that honey bee exposure to pesticides is inversely related to extent of urbanization. This pattern may be due to the lack of large-scale agriculture in urban settings and the more judicious use of pesticides in heavily populated areas. Future studies will test samples for particulate matter and heavy metals, which may be of more concern in urban areas.

EFFECTS OF HOST PLANT RESISTANCE AND INTERCROPPING ON POTATO LEAFHOPPER (*EMPOASCA FABAE*) AND ITS VULNERABILITY TO PREDATION IN ALFALFA. S.W. Vondy* and C.S. Straub, Ursinus College, Collegeville, PA (37)

ABSTRACT

Alfalfa is an important forage crop in Pennsylvania. The potato leafhopper, *Empoasca fabae*, is a key pest of alfalfa. Insecticides are effective in controlling leafhopper, but can be damaging to the environment. Host-plant resistance and intercropping are two alternative leafhopper control methods. In an open field study, we examined how potato leafhopper densities were affected by three planting treatments: leafhopper-susceptible alfalfa, leafhopper-resistant alfalfa, and leafhopper-resistant alfalfa intercropped with orchardgrass, a non-host for potato leafhoppers. We also conducted mesocosm experiments to examine how these planting treatments influence the effectiveness of the leafhoppers' primary predator, *Nabis americoferus*. In the open field, leafhopper densities were significantly higher in leafhopper-susceptible alfalfa than in the leafhopper-resistant and intercropped leafhopper-resistant alfalfa, but there was no difference between the latter two treatments. *Nabis* was found to be a more effective predator in the treatments with leafhopper-resistant alfalfa. Laboratory experiments indicated that *Nabis* was more effective in these treatments because leafhopper-resistant alfalfa caused the leafhoppers to move more, resulting in more predator-prey encounters. Increased predation appears to be reducing leafhopper abundance, making the use of leafhopper resistant alfalfa an effective means to control leafhoppers.

COVER CROPPING TO SUPPORT POLLINATORS. E.D. Treanore*, S. Fleischer, and R.C. Miller, Pennsylvania State University, State College, PA (38)

ABSTRACT

Cover cropping is an established practice in agriculture, with targeted objectives ranging from erosion control to nutrient management. Recently, floral provisioning has been gaining traction as a practice to support pollinators within agricultural environments. Here, we looked at how the benefits of both cover cropping and floral provisioning could be achieved with a two-stage planting across a growing season. As these plantings were tailored to pumpkin (*Cucurbita pepo*), species-mixtures that flowered annually, synchronized to the early and late seasonal phenology of *Bombus impatiens*, were chosen over the more commonly used perennial species in floral provisioning or single-species in cover crops.

Beginning in 2014, three locations throughout Pennsylvania were chosen for this two-stage planting. The first phase, installed in the early fall, was designed to serve as both a cover crop and as floral provisioning, with a focus on supporting early emerging *Bombus impatiens* queens the following spring. The secondary stage was designed solely to support pollinators, being planted in early July to serve as a floral resource following commercially relevant pumpkin bloom. Measures of floral density, bee visitation, seeding rate, and weed control, were taken. The relative advantages of both these plantings within the pumpkin cropping system will be discussed.

PLANTING SNAP BEANS, *PHASEOLUS VULGARIS*, ON LIGHT-REFLECTING POLYETHYLENE MULCHES PROTECTS PLANTS FROM MEXICAN BEAN BEETLE, *EPILACHNA VARIVESTIS*, FEEDING INJURY AND INCREASES POD PRODUCTION.
L.B. Nottingham* and T.P. Kuhar, Virginia Tech, Blacksburg, VA (40)

ABSTRACT

Mexican bean beetle (MBB), *Epilachna varivestis* Mulsant, is a pest of snap beans, *Phaseolus vulgaris*, and lima beans, *Phaseolus lunatus*, in eastern North America. Previous research shows that increased exposure to short wave-length light is unfavorable for MBB colonization and survival. Plastic mulches with light-reflecting surfaces, such as white and reflective silver (metalized), are known to significantly increase short wave-length light around plants. We hypothesized that beans grown on reflective plastic mulches would harbor fewer MBB, and incur less feeding injury than beans grown on non-reflective surfaces. Three field experiments were conducted in Blacksburg, Virginia: summer 2014, summer 2015, and fall 2015. 'Dragon's Tongue' snap beans were grown on beds of metalized plastic, white plastic, black plastic and bare soil. Treatment differences were measured from visual counts of all MBB life stages, ratings of foliar and pod feeding injury, and total pod weights. Short wave-length light reflectance was monitored on treatments using pyranometers. Metalized plastic reflected the greatest amount of light, followed by white, black and bare soil, respectively. Beans planted on metalized plastic consistently harbored fewer MBB adults and eggs, and exhibited less feeding injury to plants and pods than other treatments. In the two summer experiments, metalized plots had higher yields than all other treatments, with over 100 percent increases from the bare soil control; white plastic produced the highest yield in the one fall experiment. These findings suggest that growing beans on reflective plastic mulches can protect against MBB injury, while increasing marketable pod yields.

WILL THE LONG OVIPOSITOR OF *SPATHIUS GALINAE* SAVE BIG ASH?
T. Murphy*, J. Elkinton, J. Gould, and R. Van Driesche, UMASS-Amherst, Amherst, MA (41)

ABSTRACT

Agrilus plannipennis, the emerald ash borer (EAB), has spread to half the states in the US, killing millions of native ash trees. Management is focused on biocontrol with the introduction of four parasitic wasps including larval parasitoids *Spathius agrili*,

Spathius galinae, and *Tetrastichus planipennisi*. *Spathius galinae* was approved for release this year to help improve control in the northern US. Spathius agrili is not establishing in the north and *T. planipennisi* cannot oviposit in trees larger than 11.2 cm in diameter at base height (DBH). Spathius galinae's long ovipositor (4-5.3mm) is expected to help target trees with a large DBH. Climate matching indicates that there is a better fit between the native range of *S. galinae* and EAB populations in the northern US than is true for *S. agrili*. The objective of this study was to test the limits and preferences for oviposition of *S. galinae*, to help understand its potential impact on EAB management. Initial results show that *S. galinae* can oviposit across a wide range of log diameters and at a minimum can parasitize through an average valley (furrow) thickness of 4.7 mm. More data are needed using large logs to determine the upper limit of *S. galinae*.

HABITAT MODIFICATION FOR *OSTRINIA NUBILALIS* IPM IN NEW JERSEY PEPPERS. G.C. Condon* and G.C. Hamilton, Rutgers University, New Brunswick, NJ (42)

ABSTRACT

Intercropping with flowering plants can improve biological control in agricultural fields. Increased predation rates of sentinel European corn borer eggs in field grown peppers have been found with flowering intercrop use. The objective of this study was to determine the effect of 1) flowering intercrop species (fennel, coriander and dill) on anthocorid predation of sentinel European corn borer eggs, 2) intercrop plot design on anthocorid predator abundance, and 3) intercrop treatment on injury to and quality of field peppers at harvest. Two field pepper plots at the Rutgers Snyder Farm in Pittstown, NJ, were established in the summer of 2015 using a randomized mixed-block design of adjacent (5 x 10 yd) or intra-row intercrops (5 x 6 yd) subplots, with 4 replicates of 5 treatments of peppers only, peppers intercropped with fennel, coriander, dill or all 3 intercrop species. The natural abundance of anthocorid predators was determined using clear sticky traps and anthocorid predation was monitored by sentinel European corn borer egg masses. The results are discussed in terms of conservation biological control as a method to prevent economic injury in New Jersey peppers.

ADVANCES IN BLACK LIGHT MONITORING TOOLS. J.E. Cambridge* and G.C. Hamilton, Rutgers University, New Brunswick, NJ (43)

ABSTRACT

Ultraviolet light traps have been employed as a reliable insect population monitoring tool in agricultural, forest, urban, and research settings for over 70 years. As many industries strive to decrease the quantity and environmental impact of the pesticides they use, there is a growing need for more precise ways to detect and observe pest populations. In response to this, several updates and new designs for tower-type ultraviolet light traps have been tested and compared to previous models. Preliminary results suggest that improvements can be made to total catch while simultaneously reducing the cost of the materials.

ASSESSMENT OF CARFENTRAZONE FOR WEED CONTROL AND LEGUME TOLERANCE IN FORAGES. R.S. Randhawa*, M.L. Flessner, and J. Reed, Virginia Tech, Blacksburg, VA (47)

ABSTRACT

Weed control in forages is important to achieve yield, quality, and stand longevity. Legumes are a desirable component of many forage stands, but most herbicides registered for use in mixed legume-grass swards are highly injurious to legumes. Carfentrazone is a herbicide with reported safety to legumes and is registered for use in mixed-stand forages The objective of this study was to evaluate carfentrazone for weed control and legume tolerance in forages.

Multiple field experiments were conducted at Kentland Farm in; Blacksburg, VA in the summer of 2015 to evaluate carfentrazone applied alone and tank-mixed with other herbicides. Study one evaluated star-of-Bethlehem (*Ornithogalum umbellatum*) control. Treatments included carfentrazone (Aim; FMC Agricultural Products Group, Philadelphia PA) at 13.1 and 17.5 g ai ha^{-1}, carfentrazone (Aim) at 13.1 g ai ha^{-1} + 2,4-D (2,4-D LV4; Agriliance, LLC, St. Paul, MN) at 530 g ae ha^{-1}, carfentrazone (Aim) at 17.5 g ai ha^{-1} + 2,4-D (2,4-D LV4) at 530 g ae ha^{-1}, carfentrazone (Aim) applied twice, sequentially at 13.1 followed by (fb) 13.1 g ai ha^{-1}, carfentrazone (Aim) at 17.5 fb 17.5 g ai ha^{-1}, carfentrazone (Aim) at 13.1 g ai ha^{-1} + 2,4-D (2,4-D LV4) at 530 g ae ha^{-1} fb carfentrazone alone at 13.1 g ai ha^{-1}, carfentrazone (Aim) at 17.5 g ai ha^{-1} + 2,4-D (2,4-D LV4) at 530 g ae ha^{-1} fb carfentrazone alone at 17.5 g ai ha^{-1}, metsulfuron (Escort XP; DuPont, Wilmington, DE) at 16.8 g ai ha^{-1}, bromoxynil + pyrasulfotole (Huskie; Bayer Crop Science, Research Triangle Park, NC) at 244 g ai ha^{-1}, and dicamba (Clarity; BASF Crop., Research Triangle Park, NC) at 1120 g ae ha^{-1}. Sequential application treatments were made 2 weeks after initial treatment. Study two was conducted to evaluate hemp dogbane (*Apocynum cannabinum)* and yellow crownbeard (*Verbesina occidentalis)* control. Treatments included carfentrazone (Aim) at 17.5 and 35 g ai ha^{-1}, 2,4-DB (Butyrac 200; Albaugh, Inc. Ankeny, IA) at 560 g ae ha^{-1}, and metsulfuron (Ally XP; DuPont) at 12.6 g ai ha^{-1}. Other treatments were all tank mixed with carfentrazone (Aim) at 17.5 g ai ha^{-1} and included 2,4-DB (Butyrac 200) at 560 g ae ha^{-1}, metsulfuron (Ally XP) at 4.2 and 12.6 g ai ha^{-1}, 2,4-D (2,4-D LV4) at 530 g ae ha^{-1}, dicamba (Clarity) at 560 g ae ha^{-1}, dicamba + 2,4-D (Weedmaster; Nufarm Inc. Burr Ridge, IL) at 540 g ae ha^{-1}, and aminopyralid (Milestone; Dow AgroSciences LLC, Indianapolis, IN) applied at 560 g ae ha^{-1}. Study three assessed tolerance of four leguminous species: red clover (*Trifolium pratense*), alsike clover (*Trifolium hybridum*), ladino clover (*Trifolium repens*) and alfalfa (*Medicago sativa*). Treatments included carfentrazone (Aim) at 17.5 g ai ha^{-1}, metsulfuron (Escort XP) at 4.2 g ai ha^{-1}, metsulfuron + dicamba + 2,4 D (Cimarron Max; DuPont) at 1105 g ai ha^{-1}, carfentrazone (Aim) at 17.5 and 35 g ai ha^{-1} + metsulfuron (EscortXP) at 4.2 and 8.4 g ai ha^{-1} respectively; halosulfuron-methyl (Sandea; Gowan Company, Yuma, AZ) at 52.5 g ai ha^{-1}, and quinclorac (Drive XLR8; BASF Crop.) at 420 and 840 g ae ha^{-1}. All four clovers were separately treated in each treatment. All studies included a non-treated check and had four replications per treatment. Experiments utilized a randomized complete block design with the exception of the legume tolerance, which used a randomized complete split-block design. Weed control was visually evaluated relative to the non-treated check on a 0 (no control) to 100 (complete plant necrosis) scale. Visible control was assessed 2, 4, 6, 7 and 8 weeks after treatment (WAT). Legume injury was visually assessed using a similar 0 to 100 scale, with a score of 20 representing the maximum commercially acceptable injury. Data analyses were performed using SAS PROC GLM. Data were subjected to ANOVA and effects were considered significant when $P < 0.05$. Subsequently, data were also subjected to means separation using Fisher's protected LSD ($P < 0.05$).

Carfentrazone at 13.1 and 17.5 g ai ha^{-1} failed to result in commercially acceptable star-of-Bethlehem control; control was <45 percent at all rating dates. Better control was achieved from carfentrazone applied twice, sequentially at both 13.1 and 17.5 g ai ha^{-1}; control was 75 and 83 percent respectively, 6 weeks after treatment (WAT). All tank-mixed treatments resulted in >75 percent star-of-Bethlehem control 6 WAT, with no difference among the treatments. A single application of carfentrazone at either rate + 2,4-D resulted in 81 and 77 percent control, respectively that was similar to sequential carfentrazone applications. Carfentrazone at either rate + 2,4-D fb carfentrazone at 13.1 and 17.5 g ai ha^{-1} resulted in 51 and 58 percent control 2 WAT and both the treatments resulted in 89 percent control 6 WAT; this result was similar to sequential carfentrazone applied alone at either rate at both rating dates.

For hemp dogbane and yellow crownbeard control, carfentrazone applied alone at 17.5 and 35 g ai ha^{-1} resulted in <25 percent control throughout the trial. Weed height exceeded the labeled recommendations, which likely resulted in poor control observed from carfentrazone. Conversely, all tank-mixed treatments resulted > 60 percent hemp dogbane and yellow crownbeard control, except carfentrazone at 17.5 g ai ha^{-1} + metsulfuron at 4.2 g ai ha^{-1}, which resulted in <35 percent control 7 WAT. 2,4-DB applied alone resulted in 60 to 70 percent hemp dogbane control and 70 to 85 percent yellow crownbeard control, across rating dates. Metsulfuron applied alone resulted in <45 percent hemp dogbane control but resulted in >85 percent yellow crownbeard control, across rating dates. Tank-mixture of carfentrazone + metsulfuron resulted in improved hemp dogbane control relative to carfentrazone applied alone (1 versus 62 percent) 7 WAT, but the tank-mix did not improve control relative to metsulfuron alone across rating dates. Tank-mixture of carfentrazone + metsulfuron resulted in improved yellow crownbeard control relative to carfentrazone applied alone across rating dates, but the tank-mix did not improve control relative to metsulfuron alone across rating dates.

Evaluating legume tolerance, carfentrazone was clearly the safest herbicide. Carfentrazone resulted in <5 percent injury to all legume species at all rating dates except red clover 2 WAT, when 14 percent injury was observed. All other treatments resulted in ≥60 percent injury to all legume species except halosulfuron-methyl, which caused <20 percent injury in alsike and ladino clover 4 WAT. When averaged across all herbicide treatments, ladino clover was the most susceptible species and was in the most injured statistical grouping at every rating date. Ladino clover was injured >70 percent 4, 6 and 8 WAT, across herbicide treatments. Alfalfa was the only legume that recovered from treatment; maximum injury of 71 percent was observed 2 WAT, but only 39 percent injury was observed 8 WAT, across herbicide treatments. Overall these data indicate that though legume species tested are tolerant to carfentrazone, commercially acceptable control of hemp dogbane and yellow crownbeard was not achieved by carfentrazone alone. Star-of-Bethlehem was controlled by two sequential applications of carfentrazone. Future research should evaluate legume tolerance to sequential carfentrazone applications.

EVALUATING EFFICACY OF PRE HERBICIDES APPLIED EPOST IN SORGHUM. W.J. Everman, L. Vincent, and J.T. Sanders*, North Carolina State University, Raleigh, NC (48)

ABSTRACT

Grass weed control continues to be one of the greatest challenges in sorghum production in North Carolina. Several products are currently labeled for POST broadleaf weed control, however, only one product is currently labeled for POST grass control. In order to determine the best use pattern for quinclorac, a series of experiments were conducted. In 2015, a study was performed at two research stations in North Carolina which evaluated the performance of atrazine and several preemergence (PRE) herbicides paired with quinclorac at the EPOST timing in order to gauge crop safety, weed control and yield in grain sorghum. At each location, the study was arranged as a randomized complete block design with 8 treatments and 4 replications. Treatments consisted of an atrazine PRE followed by quinclorac, quinclorac + a residual herbicide, or quinclorac + a residual +atrazine applied to sorghum 10-18cm in height. The addition of atrazine caused significant but transient stunting of the crop in its early stages, but significantly improved yield. In addition, control of *Urochloa platyphylla* (broadleaf signalgrass) and *Digitaria sanguinalis* (large crabgrass) was enhanced with both the addition of residual herbicides to quinclorac and with the further addition of atrazine to those mixtures.

ASSESSING RESIDUAL BENEFITS OF POSTEMERGENT HERBICIDES USED IN HAYFIELDS. K.B. Pittman*, M.L. Flessner, and P.L. Burch, Virginia Tech, Blacksburg, VA (49)

ABSTRACT

Summer annual weeds reduce hay quantity and quality. These weeds could be controlled preemergence, but there are currently no preemergence herbicides registered for use in cool-season hayfields. Preliminary research indicates postemergence herbicides that persist in soil may provide some residual weed control. Research was conducted to evaluate control longevity from postemergence herbicides to summer annual weeds: Japanese stiltgrass (*Microstegium vimineum*), common ragweed (*Ambrosia artemisiifolia*) and common lambsquarters (*Chenopodium album*).

Multiple studies were conducted in Montgomery and Giles County in Virginia in 2015. In the first study, Japanese stiltgrass control was evaluated with three application times. Preemergence treatments were applied on April 13, 2015 and included aminopyralid (Milestone; Dow AgroSciences, LLC, Indianapolis, IN) at 0.12 kg ae ha^{-1}, aminopyralid + metsulfuron (Chaparral; Dow AgroSciences, LLC) at 0.15 kg ai ha^{-1}, aminocyclopyrachlor (DPX MAT28; DuPont Crop Protection, Wilmington, DE) at 0.18 kg ai ha^{-1}, and pendimethalin (Satellite; United Phosphorous Inc., King of Prussia, PA) at 4.49 kg ai ha^{-1}. Early-postemergence treatments were applied on May 11, 2015 and were quinclorac (Drive XLR8; BASF Ag Products, Research Triangle Park, NC) at 0.84 kg ae ha^{-1}. Postemergence treatments were applied on June 8, 2015 and were metsulfuron (Ally; DuPont Crop Protection) at 0.017 kg ai ha^{-1}, triclopyr (Remedy Ultra; Dow AgroSciences, LLC) at 0.84 kg ae ha^{-1}, glyphosate (Touchdown Total; Syngenta Crop Protection, LLC, Greensboro, NC) at 1.17 kg ae ha^{-1}, aminopyralid (Milestone) at 0.12 kg ae ha^{-1}, aminopyralid + metsulfuron (Chaparral) at 0.15 kg ai ha^{-1}, and aminocyclopyrachlor (DPX MAT28) at 0.18 kg ai ha^{-1}. The second trial evaluated common ragweed and common lambsquarters control over time. Treatments were applied on May 5, 2015 and included 2,4-D + aminopyralid (GrazonNext HL; Dow AgroSciences, LLC) at 0.77 kg ae ha^{-1}, 2,4-D + picloram (Grazon P+D; Dow AgroSciences, LLC) at 0.56 kg ae ha^{-1}, aminocyclopyrachlor (DPX MAT28) at 0.07 kg ha^{-1}, aminopyralid (Milestone) at 0.12 kg ae ha^{-1}, picloram + fluroxypyr (Surmount, Dow AgroSciences, LLC) at 0.28 kg ae ha^{-1}, clopyralid (Transline, Dow AgroSciences, LLC) at 0.63 kg ae ha^{-1}, aminopyralid + metsulfuron (Chaparral) at 0.011 kg ae ha^{-1}, 2,4-D amine (DuPont Crop Protection) at 1.12 kg ae ha^{-1}, saflufenacil (Sharpen; BASF Ag Products) at 0.05 kg ai ha^{-1} and pendimethalin (Satellite) at 1.06 kg ai ha^{-1}. The third study also evaluated common lambsquarters and common ragweed control over time. Treatments were 2,4-D + aminopyralid (GrazonNext HL) at 0.79 kg ae ha^{-1} and 0.63 kg ae ha^{-1}, aminopyralid + metsulfuron (Chaparral) at 0.11 kg ae ha^{-1}, 2,4-D + picloram (Grazon P+D) at 0.72 kg ae ha^{-1}, 2,4-D (DuPont Crop Protection) at 1.12 kg ae ha^{-1}, and 2,4-D + dicamba (Weedmaster; Nufarm Agricultural Products, Alsip, IL) at 1.09 kg ae ha^{-1}. All treatments were applied at two different application timings: May 5, 2015 and June 19, 2015, respectively. All experiments included a nontreated check, utilized a randomized complete block design with four replications and were applied at 140 L ha^{-1}. Control was visually evaluated relative to the nontreated check on a 0 (no control) to 100 (complete plant necrosis) scale. Visible control was assessed 2, 4, 6, 7, 8, 10, 12 and 15 weeks after treatment (WAT) for Japanese stiltgrass control and 1, 2, 4, and 8 WAT for common lambsquarters and common ragweed. Injury was assessed on orchardgrass (*Dactylis glomerata*) in the second and third study relative to the nontreated check on a scale of 0 (no injury) to 100 (complete plant necrosis) 1, 2, 4, and 8 WAT. Data were subjected to ANOVA and effects were considered significant when $P<0.05$ followed by means separation using Fisher's protected LSD ($P<0.05$).

For Japanese stiltgrass control, aminopyralid + metsulfuron and aminocyclopyrachlor resulted in the best control (81 and 68 percent, respectively) 2 weeks after the early-postemergence application. However, all preemergence and early-postemergence treatments resulted in <50 percent control at the final rating (15 weeks after initial treatment). Glyphosate, aminopyralid, aminopyralid + metsulfuron, and aminocyclopyrachlor resulted in the best control (99, 83, 82, and 94 percent, respectively) 8 weeks after postemergence treatments. In the second study, all herbicides tested resulted in ≥75 percent common lambsquarters control 4 WAT except saflufenacil, clopyralid, and 2,4-D amine. Common lambsquarters control was ≥75 percent for all treatments 8 WAT. For common ragweed, control was ≥90 percent 4 WAT for all treatments except pendimethalin and 2,4-D amine, which resulted in 0 and

10 percent control, respectively. Similarly, all herbicides tested resulted in complete (100 percent) common ragweed control 8 WAT, except pendimethalin and 2,4-D, which resulted in 0 and 58 percent control, respectively. Saflufenacil resulted in 25 percent orchardgrass injury 1 WAT, characterized by necrotic spots, but a complete recovery from injury was observed 4 WAT. In the third study, all treatments resulted in ≥51 percent common ragweed and common lambsquarters control 8 weeks after the first application (2 weeks after the second application). All treatments resulted in ≥88 percent control 4 weeks after the second application except for 2,4-D, which resulted in 0 percent common lambsquarters control and 4 percent common ragweed control. 2,4-D has a short soil residual (<7 day half-life). 2,4-D resulted in initial weed control in studies two and three, but no control later in the season most likely due to subsequent germination. Conversely, herbicides with a long soil residual (≥30 day half-life) generally resulted in weed control throughout the season. Therefore these studies indicate that herbicides with significant residual soil activity can achieve season-long annual weed control in hayfields when applied early-postemergence. Future research is necessary to corroborate preemergence control findings, assess other weed species, and determine optimum application timing.

A LITERARY REVIEW EVALUATING THE NEED FOR MORE EDUCATION AND INCREASED MONITORING OF THE URBAN PEST CONTROL INDUSTRY. D.M. Bailey* and D. Pfeiffer, Virginia Tech, Blacksburg, VA (50)

ABSTRACT

Since states were required by the Environmental Protection Agency (1970) to regulate training, test and license those applying pesticides; the majority of focus has been directed towards private applicators working within farm settings. Urban pesticide applicators are also required to train, test , become licensed and adhere to both federal and state laws, but scrutiny is much easier to avoid because urban pesticide applicators' work is often concealed by walls, varied times of application and an unfair ratio to state inspectors. While state agencies have allotted budgets for extension agents to design programs and travel for the sole purpose to educate and recertify applicators, these resources are often heavily geared towards the agriculture industry. As well, the research being conducted concerning pesticide safety, level of understanding, environmental awareness, personal protection equipment and other topics related to insect biology and habits is mainly studied from the aspect of a farmers for the farmworkers. Understandably focus, training and budgets should be allocated for individuals handling the food that feeds the world but there needs be an equal amount of resources dedicated to the training of applicators serving metropolitan communities and entrusted with dispersing chemical solutions into personal, professional and public environments.

VIRGINIA CROP WEED MANAGEMENT SURVEY. S.C. Haring* and M.L. Flessner, Virginia Tech, Blacksburg, VA (51)

ABSTRACT

Herbicide resistance (HR) is a growing problem in crop production across Virginia, and statewide information regarding weed management practices is largely anecdotal. This information is necessary to identify high-impact Extension education and outreach areas. A survey was conducted to gain information regarding problematic and HR weed species, weed management practices, and costs related to HR directly from stakeholders in Virginia.

Surveys were conducted in-person (on paper) and online. Recruitment of in-person subjects occurred at county and regional agriculture conferences organized by Virginia Cooperative Extension and the Virginia Crop Production Association. Online subject recruitment occurred through email lists, including email lists maintained by Virginia Cooperative Extension, Virginia Soybean Producers Association, and Virginia Grain Producers Association, among others. These populations were selected because, together, they comprise a representative sample of stakeholders in crop weed management from Virginia. Surveys were administered between February and April 2015. Online surveys were administered through Qualtrics Research Suite software (Qualtrics, LLC, Provo, UT). Paper survey data was entered into this software, which was used to compile survey results. The survey was completed by 97 growers, who collectively farm about 10 percent of corn, soybean, and small grains acreage in Virginia. Out of 87 responses, 53 percent of growers used some mixture of glyphosate resistant, glufosinate resistant, and conventional crops. Glyphosate resistant and conventional crops were used exclusively by only 30 percent and 17 percent of growers, respectively.

Overall, 53 growers identified HR weeds on their farms, representing 55 percent of survey participants, which was an increase from 44 percent in a 2011 survey. Of these 53 growers, 30 percent identified glyphosate resistant palmer amaranth (*Amaranthus palmeri*), 28 percent identified glyphosate resistant horseweed (*Conyza canadensis*), and 17 percent identified glyphosate resistant common ragweed (*Ambrosia artemisiifolia*) to be present on their farms. However, 26 percent of reported HR cases involved weed species not confirmed to be HR in Virginia, indicating that misidentification of HR is a major issue. Corn, soybean, and small grains growers estimated that HR weeds have led to an average increase in weed management costs of $10.00, $9.67, and $5.50 per acre, respectively. These increases represent 18 percent, 21 percent, and 18 percent of all estimated per acre weed management costs, respectively. Increased costs could be attributed to changes in cultural practices to manage HR weeds. The use of multiple modes of herbicidal action had been adopted by 23 percent of growers in response to HR weeds, while 33 percent growers reported having always used this practice. Additionally, 21 percent of growers began to rotate modes of action in response to HR, while 24 percent had always used this practice. Other management practices have seen no change in use in response to HR. Crop destruction techniques, for example, have seen low levels of adoption, with 45 percent of growers having never used the practice. Conversely, equipment cleaning, crop scouting, crop rotation, and the use of preemergent herbicides are widespread weed management practices that had been regularly practiced by 45 percent, 45 percent, 35 percent, and 31 percent of growers, respectively. These standard practices were used to combat the most troublesome non-HR weeds, which growers also identified. Morningglory species (*Ipomoea* spp.) and redroot pigweed (*Amaranthus retroflexus*) were identified by 22 percent and 16 percent, respectively, of 45 corn producers. Horseweed and redroot pigweed were each identified by 16 percent of 62 soybean producers. Italian ryegrass (*Lolium multiflorum*) and henbit (*Lamium amplexicaule*) were identified by 31 percent and 14 percent, respectively, of 58 small grains producers. The diversity of growers' response and needs regarding HR weeds has important implications and can provide insight for the development of Extension education and outreach programs.

MOLECULAR ANALYSIS OF GLYPHOSATE RESISTANCE IN GIANT RAGWEED (*AMBROSIA TRIFIDA* L.). K. Segobye*, K.R. Padmanabhan, B. Schulz, M. Gribskov, and S.C. Weller, University of Maryland, College Park, MD (52)

ABSTRACT

Giant ragweed (*Ambrosia trifida* L.) is one of the most competitive weeds found in corn and soybean cropping systems, leading to huge production losses in the Corn Belt and the Southern United States. The use of glyphosate and roundup-ready crops provided a tool to control giant ragweed but its management is becoming more difficult with the recent occurrence of glyphosate resistance. Our research investigates the plant biology and molecular mechanism of glyphosate resistance in giant ragweed biotypes. The goal of our research is to determine the mechanism of hypersensitive-like reactions in glyphosate resistant giant ragweed and determine the genes responsible for glyphosate resistance.

We hypothesize that the basis of resistance is reduced translocation of glyphosate, which is initiated by a signaling mechanism for resistance in resistant plants. Glyphosate resistant plants receive glyphosate as a "pseudo pathogen" and react by employing the present pathogen response system of plants. This includes cell responses that normally result in hypersensitive cell death similar to a response to pathogen attack. To test this we conducted experiments with resistant and susceptible biotypes to assess responses to different doses of glyphosate. The two biotypes were compared for response to glyphosate at rates of 0.5 X, 1X, 2X, 4X and 8 X application rates (1X= Field rate=0.84kg ae/ha). Both biotypes were also studied to investigate if the production of reactive oxygen species (ROS) - typical components that are involved in the regulation of pathogen response system - is involved in rapid necrosis of mature leaf tissues of glyphosate resistant giant ragweed plants following glyphosate treatment. We also analyzed the two biotypes for glyphosate response at the molecular level by assessing their reaction and analyzing the total transcriptome of glyphosate treated and untreated plants.

Glyphosate resistant plants show a unique response when treated with the herbicide, exhibiting initial rapid necrosis of mature leaves within 12 hours of treatment. Glyphosate resistant plants do not die from a glyphosate treatment but resume normal-growth from axillary meristems and produce viable seeds. Glyphosate susceptible plants do not exhibit rapid leaf necrosis but their leaves become chlorotic, then necrotic and plants die over a 2-3 week period. The results also revealed that responses to glyphosate in resistant giant ragweed plants lead to reprogramming of plant development and changes in plant architecture. The time course study demonstrated that hydrogen peroxide (H_2O_2) accumulates in mature leaves of resistant giant ragweed as early as 2 hours after glyphosate treatment visually detected after 3, 3'-diaminobenzidine (DAB) staining procedures. We have identified a list of genes that were differentially expressed between the two biotypes as the first step in identifying genes responsible for the glyphosate resistance observed and help in understanding the mechanism of giant ragweed resistance to glyphosate.

SCREENING FOR HERBICIDE RESISTANCE IN OHIO JOHNSONGRASS POPULATIONS. A. Lamb*, The Ohio State University, Columbus, OH (53)

ABSTRACT

Johnsongrass (*Sorghum halepense*) is a common perennial noxious weed found in crop production fields in Ohio. It has the ability to develop glyphosate resistance and has done so in both Argentina and Arkansas. Two Ohio johnsongrass populations were suspected to have resistance to glyphosate based on inadequate late-season control in glyphosate-resistant soybeans. A preliminary greenhouse study indicated the possibility of such resistance, and was followed by additional greenhouse and field studies. Response of these populations was compared with a known sensitive population in the greenhouse, using glyphosate rates of 0.84 and 3.3 kg ae/ha. Control of all populations exceeded 95 percent at both rates. A field study was conducted at one of the suspect sites using the same glyphosate rates, and also included single and multiple glyphosate treatments, a clethodim treatment and a non-treated control. All treatments at the field site provided complete johnsongrass control. Based on these results, the presence of glyphosate resistance in these populations was not confirmed.

CROP DENSITY EFFECTS ON INTERSEEDED COVER CROP PRODUCTIVITY AND WEED SEED PREDATION. C.Z. Youngerman*, W. Curran, S. Wayman, and M.R. Ryan, Cornell University, Ithaca, NY (54)

ABSTRACT

Corn and soybean farmers face major challenges with weed suppression and soil fertility during the transition to certified organic production. We established an experiment in NY and PA in 2015 to evaluate the effects of crop density on the performance of interseeded cover crops, weed suppression, and weed seed predation during the transition to certified organic production. Corn and soybean were planted at five rates ranging from approximately 50-100k and 125-620k seeds/ha, respectively. Crop density treatments were arranged in a randomized complete block design with four blocks. A cover crop polyculture (51 percent cereal rye cv. 'Huron', 25 percent annual ryegrass cv. 'KB Supreme', 14 percent hairy vetch cv. 'VNS', and 10 percent medium red clover cv. 'VNS') was interseeded into corn and soybean at 66 kg seed/ha after the last inter-row cultivation. Light transmittance through the crop canopy, invertebrate activity diversity, and weed seed predation were measured at 20, 40, and 60 days after interseeding. Cover crop and weed biomass were sampled at crop maturity and crop grain yield was quantified at harvest. Weed seed predation was quantified using seed arenas with and without invertebrate exclosures. As we were interested in the impact of interseeded cover crops on weed seed predation, we included cover crop seeds in addition to weed seeds in our assessment. Arenas included 25 seeds each of *Setaria faberi*, *Amaranthus powellii*, *Secale cereale,* and *Trifolium pratense*. Preliminary results show crop yield increases with planting rate and that cover crop biomass production and seed predation were similar across all crop densities. Open seed arenas had more than double the predation rates than closed arenas. Invertebrate seed predators preferred *S. faberi* and *T. pratense* over *A. powellii* and *S. cereale*, whereas *S. cereale* was the consumed more than other species in the open arenas. Seed predation of interseeded cover crops can potentially limit their establishment and the weed suppression and soil fertility benefits they provide.

INTEGRATING COVER CROPS INTO MID-ATLANTIC NO-TILL GRAIN SYSTEMS TO DIVERSIFY HERBICIDE RESISTANCE MANAGEMENT. J.M. Bunchek*, Penn State -- Dept. of Plant Sciences, State College, PA (55)

ABSTRACT

In order to combat a rapid increase in glyphosate-resistant horseweed and pigweed species, we are assessing the integration of cover crops to diversify herbicide-resistant weed management in the Mid-Atlantic United States. This region is comprised heavily of dairy farms, where conservation-tillage practices are being promoted to preserve soil health and restore the Chesapeake Bay Watershed. Thus, all field experiments in this first-year study integrate no-till farming. The main objectives are (1) to test the effects of cover crop functional traits on winter- and summer-annual weed population dynamics; (2) to quantify the efficacy of cover crop and herbicide strategies on weed density, control, and fecundity; and (3) to evaluate trade-offs associated with ecologically-based weed management by assessing how cover crop strategies contribute to herbicide resistance management, ecosystem services, and grower priorities. To address these objectives, we have initiated three field experiments at Penn State's Russell E. Larson Agricultural Research Center outside State College, PA, and the University of Delaware's Carvel Research and Education Center in Georgetown, DE. The protocols for these three field experiments derive from a preliminary experiment initially conducted in 2014-2015, which concluded that fall-planted cover crops have the ability to significantly inhibit horseweed populations and therefore reduce herbicide selection pressure. Further, winter-kill cover crops produced more ground cover at 5 weeks after planting (WAP) than winter-hardy cover crops, providing greater horseweed suppression. The three experiments currently in place include one cover crop species (cereal rye or spring oats) or mixtures of two cover crop species (cereal rye, spring oats, hairy vetch, forage radish, and crimson clover), and each mixture consists of two winter-kill species, two winter-hardy species, or one winter-kill and one winter-hardy species. Cover crops were seeded into fields in 2015 and will be succeeded by soybean or corn.

The cover crop mixtures in the first experiment will be terminated at two dates each in the spring and analyzed for suppression of horseweed and pigweed species. Fall data collection included cover crop ground cover and horseweed population counts taken at 5 WAP and 10 WAP, and cover crop biomass (kg ha^{-1}) was taken at 10 WAP within a 0.5 m^{-2} area per plot. Spring data collection will consist of percent cover, horseweed and pigweed population counts, and biomass collection taken at termination. We can hypothesize that delaying cover crop termination by two weeks will significantly reduce horseweed and pigweed species size and density during the growing season. The second and third experiments study the impact of cover crop residue along with herbicide combinations (burndown only, burndown + PRE, burndown + POST, and burndown + PRE + POST). Both experiments will be assessed for weed suppression efficacy and cover crop performance by assessing horseweed and pigweed species population counts within a 0.25 m^{-2} microplot before and 10 days after termination for each herbicide treatment. We can hypothesize that for both experiments, burndown + PRE + POST will control horseweed and pigweed species significantly greater than burndown only, and burndown + PRE and burndown + POST will target early- and late-growing season horseweed and pigweed species control, respectively.

ALTERNATIVES FOR REDUCING TILLAGE IN ORGANIC ANNUAL GRAIN SYSTEMS: IMPLICATIONS FOR WEED MANAGEMENT. R. Champagne*, W.S. Curran, C. Keene, J. Wallace, and T. Mazzone, Penn State University, State College, PA (56)

ABSTRACT

One of the biggest factors affecting crop performance is interference by weedy species. Tillage and use of herbicides are common ways to combat this issue; however, weeds are adapting and evolving herbicide resistance, prompting farmers to use even more herbicides and potentially resort to more tillage. Costs associated with herbicides and potential erosion from soil disturbance are causing farmers to rethink approaches to fighting weeds through alternative methods. We are addressing some of these concerns by investigating alternative approaches to weed management in an organic grain system that is focused on crop rotation, cover cropping, and reduced tillage. Our previous organic research project in general demonstrated effective control of weeds below economic thresholds; however, it also demonstrated agronomic challenges such as inconsistent termination of cover crops with the roller-crimper, inconsistent establishment of cash crops in high residue, and issues with a narrow timeline for establishing cover and cash crops following both corn and soybean. To address these concerns, we have refined these systems and developed alternative strategies such as relay planting cover crops using interseeder technology, and companion seeding of cover crops in spring into a winter cereal grain. Weed seedbank dynamics, growth, and fecundity are being studied, along with cover crop and cash crop performance. It is our goal to improve cover crop and reduced tillage strategies in organic grain systems in the northeastern U.S. and to use these results to reduce grower-defined production pressures.

DEVELOPMENT OF SPECIES-SPECIFIC PCR TESTS FOR THE DETECTION AND QUANTIFICATION OF *MELOIDOGYNE HAPLA*. A.M. Gorny*, X. Wang, and S.J. Pethybridge, Cornell University, Geneva, NY (57)

ABSTRACT

The Northern root-knot nematode, *Meloidogyne hapla,* is a significant soilborne pathogen of potatoes and other vegetables grown in the United States and is common within New York State. Effective management of plant-parasitic nematodes relies on the accurate identification and timely quantification of pathogen populations prior to planting. Prediction of damage from soilborne diseases may be substantially improved by the provision and adoption of pre-plant soil tests, which utilize quantitative PCR (qPCR) techniques to target DNA sequences specific to the pathogen of interest. Here, PCR primers designed to target unique regions of an effector gene *16D10* in *M. hapla* were developed and assayed for specificity against thirteen plant-parasitic nematodes using qualitative PCR. The primers developed in this study were shown to be highly specific to *M. hapla.* These primers will be utilized in future qPCR tests for quantification of nematode population densities in soil and to assess the risk of crop damage or loss due to *M. hapla*.

POTENTIAL USE OF GRAS COMPOUNDS FOR THE CONTROL OF LETTUCE BACTERIAL LEAF SPOT AND VARNISH SPOT. M. Delisle-Houde*, V. Toussaint, A. Gosselin, and R.J. Tweddell, Université Laval, Quebec, QC (58)

ABSTRACT

Over the past decades, numerous studies have reported that several GRAS (Generally Recognized as Safe) compounds with antimicrobial activity show potential to control plant diseases. Bacterial leaf spot (*Xanthomonas campestris* pv. *vitians*) and varnish spot (*Pseudomonas cichorii*) are economically important diseases of commercial lettuce (*Lactuca sativa*) production in Canada and no chemicals are registered to control bacterial diseases in this crop. Consequently, control of bacterial leaf spot and varnish spot relies on cultural practices including good soil drainage and air movement through crop canopy. However, cultural practices are not always sufficient to effectively control the diseases. The objectives of the study were (1) to test different GRAS compounds (potassium sorbate, sodium benzoate, sodium bicarbonate, sodium carbonate, and sodium metabisulfite) for their antibacterial activity against *X. campestris* pv. *vitians* and *P. cichorii* and (2) to test their efficacy to control varnish spot. Determination of antibacterial activity of the salts was carried out using standard procedures in liquid medium. Efficacy of salts to control varnish spot was determined on lettuce plants grown in greenhouse. The results obtained showed that sodium bicarbonate, sodium carbonate, and sodium metabisulfite were particularly toxic to both bacteria while potassium sorbate and sodium benzoate were toxic to a lesser extent. Applied on lettuce plants inoculated with *P. cichorii*, only potassium sorbate and sodium benzoate reduced significantly varnish spot development but did not allow complete control of the disease.

EFFICACY AND ECONOMICS OF FUNGICIDES IN MID-ATLANTIC SMALL GRAIN PRODUCTION SYSTEMS, 2015. P.N. Sylvester* and N.M. Kleczewski, University of Delaware, Newark, DE (59)

ABSTRACT

Fungicides are commonly used for management of fungal diseases of wheat in the Mid-Atlantic region. The most commonly occurring disease issue in the region is Leaf Blotch Complex (LBC), caused by a set of residue borne, fungal pathogens including *Pyrenophora tritici-repentis*, *Septoria tritici*, and *Stagonospora nodorum*. Standard foliar fungicide applications occur around flag leaf emergence [Feekes growth stage (FGS) 8]. However, over the last several years fungicide programs including applications prior to FGS 8 at greenup (FGS 5), at flowering (FGS 10.5.1), and two pass programs (FGS 5 + 8 or FGS 10.5.1) have been implemented. There is little regional data pertaining to the efficacy of these programs and their impact on yield. In addition, programs differ in application cost, which will affect their grower profitability. Experiments were conducted at four sites in Delaware and Maryland to compare the new programs to standard Feekes 8 programs for efficacy, yield response, and potential profitability. Each site consisted of a randomized complete block design with six replications and 13 treatments. Treatments consisted of five commonly used fungicides applied at FGS 8, FGS 10.5.1, and a combination of FGS 5 + FGS 8 or 10.5.1. Foliar disease severity was measured over time and yields obtained. Data from each location were combined and analyzed using a random effects mixed model. Disease pressure from LBC was present at all sites to different degrees, ranging from 3.4 percent to 12.6 percent severity on the flag leaf in the untreated controls. Relative to untreated controls, all fungicide programs significantly reduced foliar disease by 75 percent to 94 percent and improved yield by 4.3 percent to 6.7 percent. Across timings, no differences between standard FGS 8 and new fungicide programs were noted. Net returns from 2015 indicated that 9 of the 13 tested programs returned a profit using 2015 commodity prices. Data from 2015 indicate that fungicide use can be profitable under Mid-Atlantic conditions, but additional fungicide applications at FGS 5 may be unnecessary. In addition, preliminary data indicate that single fungicide applications at FGS 10.5.1 may be as efficacious and protect yield as well as standard FGS 8 applications. The study will be replicated in 2016 on an additional seven sites across DE, MD, VA, and PA to build a more robust dataset and generate models and decision tools to assist growers in making profitable fungicide application decisions to small grains.

FRIEND OR FOE: *FUSARIUM OXYSPORUM* ISOLATES RESULT IN CONTRASTING MOLECULAR AND PHYSIOLOGICAL RESPONSES IN *ARABIDOPSIS THALIANA*. K. Vescio*, University of Massachusetts, Amherst, MA (60)

ABSTRACT

Fusarium oxysporum (Fo) is a soil-dwelling fungal pathogen that results in vascular wilt diseases on a broad range of plants including agricultural crops like cucurbits, legumes, and banana as well as the model plant *Arabidopsis thaliana*. This research seeks to understand the molecular interaction of the plant-microbe relationship between

A. thaliana interacting with either a pathogenic or non-pathogenic Fo formae specialis. The communication between the plant and fungus on the transcript level has been captured by timecourse RNA-sequencing, and revealed candidate plant and fungal genes for characterization.

IDENTIFICATION OF CANDIDATE GENES FOR STUDY OF THE SWEET BASIL HOST RESPONSE TO INFECTION BY *PERONOSPORA BELBAHRII*. K.S. Allen*, L. Guo, L. Ma, and R.L. Wick, University of Massachusetts, Amherst, MA (61)

ABSTRACT

Basil downy mildew is caused by *Peronospora belbahrii*, an oomycete obligate parasite infecting the mesophyll tissue of sweet basil (*Ocimum basilicum*). In U.S. regions affected by the disease, growers consistently reported 100 percent crop loss, with estimated economic losses in the tens of millions of dollars. Chemical controls for basil downy mildew are limited, variably effective, and risk increased pathogen resistance. Working with previously-generated RNA-seq data has revealed basil and *Peronospora* transcripts that showed significant levels of differential expression post-infection. These results are being validated with quantitative reverse transcription polymerase chain reaction (qRT-PCR). Differentially-regulated transcripts that are validated in this experiment will be utilized to probe the basil host response to infection, including defining important time-points in the infection process that could be targeted for more effective screening or treatment in future research.

EVALUATING CULTIVAR SUSCEPTIBILITY AS A POTENTIAL MANAGEMENT TOOL FOR CENTER ROT OF ONION. J.D. Mazzone*, M.A. Mansfield, and B.K. Gugino, The Pennsylvania State University, University Park, PA (62)

ABSTRACT

Onion growers in Pennsylvania are continually challenged by in-field and post-harvest yield losses due to center rot disease caused by the bacterial pathogens *Pantoea ananatis* and *Pantoea agglomerans*. Identifying resistant or tolerant cultivars would mitigate some of these losses. In 2015, onion cultivars Sedona, Great Western, Aruba, Ovation, BGS 280 F1, BGS 300 F1 Blush (Blush), Red Sky, Spanish Medallion, Expression and Candy were evaluated for their susceptibility to center rot as well as horticultural characteristics including neck diameter, bulb diameter and leaf length. Treatments were arranged in a randomized complete block design and plots were split by inoculation status. One inner leaf of select plants was toothpick inoculated with a mixed culture of *P. ananatis* and *P. agglomerans*. At harvest, plants were topped and graded for size, marketability and disease incidence. A sub-sample of asymptomatic onions was evaluated for soluble solids (percent sugar), pungency (pyruvic acid) and post-harvest disease incidence after 3-months in storage. Marketable yield (≥ 7.6 cm bulb diameter) ranged between 33.6 and 109.8 kg/ 73.1 m row length harvested. Cvs. Expression and Great Western had the highest marketable yield, while cvs. Blush and

Sedona had the lowest ($P<0.0001$). Center rot disease incidence significantly differed between cultivars at harvest ($P<0.001$), but not after 3-months in storage ($P=0.345$). All cultivars evaluated met the minimum soluble solids and pungency criteria to be considered for the Pennsylvania Simply Sweet Onion Program. Of these, disease incidence was lower in 5 cultivars compared to the commercial standard cv. Candy. Cv. Aruba had the highest disease incidence at harvest (17 percent) while cvs. Blush and Red Sky had the lowest (1 percent each). Based on their horticultural characteristics and disease susceptibility, several cultivars are suitable for the Pennsylvania Simply Sweet Onion Program. Cultivar evaluations will be ongoing in the 2016 season where trials will be held in multiple locations to further determine cultivars that show promise in sweet onion production.

NEWSS STUDENT CONTEST - ORAL PRESENTATIONS (I)

GREENHOUSE EVALUATION OF SPRAY ADJUVANTS AND FERTILIZER ADDITIVES FOR GRASS WEED MANAGEMENT WITH QUINCLORAC.

L. Vincent, W.J. Everman, and J. Copeland*, North Carolina State University, Raleigh, NC (63)

ABSTRACT

Historically, postemergence grass weed management in grain sorghum (*Sorghum bicolor*) has been difficult because of limited options. In 2013, BASF introduced 'Facet L' a quinclorac product to provide grass weed control in grain sorghum. The 'Facet L' label requires use of crop oil concentrate (COC), methylated seed oil (MSO), or a nonionic surfactant (NIS) while the addition of fertilizer additives, urea ammonium nitrate (UAN) or ammonium sulfate (AMS), are optional. Given that grass weed control is critical for grain sorghum production, effective tank-mixes and spray additives were evaluated to inform producers of tank-mix options that are most valuable for grass weed management.

Studies were conducted at the Method Greenhouse Facility in Raleigh, NC in 2015 to evaluate the impact of various combinations of spray adjuvants and fertilizer additives on six common grass weed species in grain sorghum production in North Carolina. Grass weed species included large crabgrass (*Digitaria sanguinalis*), goosegrass (*Eleusine indica*), broadleaf signalgrass (*Urochloa platyphylla*), fall panicum (*Panicum dichotomiflorum*), texas millet (*Urochloa texana*), and crowfootgrass (*Dactyloctenium aegytium*). Quinclorac was applied at 0.29 kg ae ha^{-1} alone and in combination with adjuvant treatments and fertilizer additives that included COC at 2.34 L ha^{-1}, MSO at 2.34 L ha^{-1}, NIS at 0.35 L ha^{-1}, UAN at 2.34 L ha^{-1}, and AMS at 1.43 kg ha^{-1}, respectively as well as an untreated check. Experiments were conducted using a factorial arrangement of treatments within a randomized complete block design, with three factors being species, adjuvant, and fertilizer additive. All data were subjected to analysis of variance and means were separated using Fisher's Protected LSD at $p= 0.05$.

Broadleaf signalgrass treated with quinclorac, regardless of adjuvant, resulted in ≥ 95 percent visual control 14 DAT. MSO treatments resulted in significantly greater visual control for broadleaf signalgrass, large crabgrass, and fall panicum when compared to

COC or NIS. Visual control 7 and 14 DAT was minimal for texas millet (<5 percent), crowfootgrass (0 percent), and goosegrass (0 percent) regardless of adjuvant combination. Height reductions at 14 DAT indicated that the combination of quinclorac and NIS resulted in a significantly less height reduction for weed species fall panicum (56 percent) and large crabgrass (72 percent) when compared to MSO (92-98 percent) or COC (90 percent). Dry weight reduction was significantly affected by weed species. Dry weight reductions were significantly greater for broadleaf signalgrass (97 percent) and large crabgrass (88 percent) when compared to fall panicum (75 percent) and texas millet (44 percent). Dry weight reductions for crowfootgrass (0 percent) and goosegrass (0 percent) were significantly less than all other weed species. Trends in texas millet, crowfootgrass and goosegrass control provide that these species may be tolerant to quinclorac. When using quinclorac to control susceptible weed species, choice of adjuvants will affect control of target weed species.

IMPACT OF PRE-PLANT NITROGEN RATES ON WHEAT FOLLOWING SORGHUM AND CORN. M.K. Bansal*, North Carolina State University, Raleigh, NC (65)

ABSTRACT

North Carolina sorghum production has renewed interest in recent years as regional grain demands increased which lead swine producer to offer a competitive sorghum grain price. Sorghum can be a good alternative for corn in rotation with wheat and provides several advantages over soybean where nematodes or weeds are a concern. Sorghum has the ability to tolerate hot dry weather, a condition that can be challenging for corn in the summer. However, with the advantages, sorghum has some disadvantages as well when used in rotation with wheat. Grain sorghum is known to produce allelochemical called 'Sorgoleone' which can have negative impact on the following wheat crop. Sorghum residue when incorporated in soil can immobilize N, making it less available to following wheat.

Experiments were conducted in 2013-14 at Rocky Mount and 2014-15 at Rocky Mount and Kinston (two locations), North Carolina to evaluate the effect of different rates of pre-plant nitrogen (15, 30, 45, and 60 lbs per acre) applied to wheat following different hybrids either sorghum (DKS 53-67, P83P17) or corn (DKC 60-67) on wheat yield. In 2013-14, there was no significant effect of pre-plant nitrogen on wheat yield. There was significant effect of hybrids on wheat yield. Wheat yield was not significantly different when planted after either DKC 60-67 or DKS 53-67. Yield was significantly different when planted after DKC 60-67 and P83P17. This reduction in wheat yield after P83P17 compared to corn could be due to sorghum allelopathy. In 2014-15, there was no significant effect of different hybrids on wheat yield at all three locations. Pre-plant nitrogen had significant effect only at one location in Kinston. Results suggests that wheat yield is not affected when planted after sorghum (DKS 53-67) compared to corn (DKC 60-67). There was no significant effect of pre-plant nitrogen at Rocky Mount in both years.

COVER CROP AND WEED MANAGEMENT IN AN INTER-SEEDED COVER CROP PLUS REDUCED-RATE HERBICIDE SYSTEM IN WIDE-ROW VEGETABLES. V. Bhaskar*, R. Bellinder, M. Walter, and A. DiTommaso, Cornell University, Ithaca, NY (66)

ABSTRACT

Herbicides and living mulches are two important weed management tools. Herbicides enhance crop yields but their use can raise environmental concerns. Living mulches have positive agroecosystem effects but have not been widely adopted due to concerns of intercrop-cash crop competition and poor weed suppression. The main objective of this study is to assess whether inter-seeded cover cropping can be feasible in combination with herbicide applications. It is hypothesized that herbicides and living mulches in this system are complimentary and will address their individual drawbacks to provide effective weed control and acceptable crop yields. Due to the many benefits of having living mulches, herbicides must be evaluated as a tool to make these systems more viable. Reduced herbicide rates in addition to improving weed control help manage cover crop growth, so that competition with the main crop is minimized. If legumes are used as living mulches, nitrogen use efficiency can be improved and competition with the main crop for N can be reduced. For this system to be workable, cover crops, herbicides, and application rates have to be selected carefully. A preliminary trial was conducted in 2014 in Freeville, NY using sesbania (*Sesbania sesban*) and sunnhemp (*Crotalaria juncea*) as living mulches, in a fresh-market tomato (Mountain Fresh F1) crop. This trial was set up as a randomized complete block design with 4 replicates. Tomato plants were transplanted with a row spacing of 1.22 m and plant-to-plant spacing of 0.46 m. Three rows of cover crops were sown between tomato rows so that cover crop rows were spaced 0.23 m apart, and 0.38 m from the tomato row on either side. Metribuzin, rimsulfuron and halosulfuron were used at two rates; control plots were hand-weeded. Results from 2014 were promising; with sunnhemp plus metribuzin plots showing <10 percent weed cover and tomato yields higher than the control treatment. In 2015, a similar trial was set up, using combinations of the herbicides used in 2014 (plus fomesafen); an untreated cover crop check, with no herbicide applications and a weedy check were also included. Herbicide combinations consisted of two applications of two herbicides. One of the herbicides was selected because the cover crop was sensitive to it when applied post-emergent. The cover crop was less sensitive to the other herbicide but this herbicide had residual activity. A sole cover crop trial was also conducted during the second year to assess the performance of additional herbicides (imazethapyr and s-metolachlor). Data collected included cover crop and weed ground cover, density and aboveground biomass, cover crop height, and tomato yield. Tomato leaf nutrient content was also determined. In 2015, tomato yield in herbicide-treated plots did not differ from the hand weeded control plots, but yield was higher than the untreated and weedy check plots ($p = 0.0027$). There was a strong positive correlation between the amount of cover crop biomass produced and tomato yield ($p = 0.0075$). In the sole cover crop trial, weed biomass from the weedy check (11.7 tons ha^{-1}) was higher than from all other herbicide treatments, of which the highest was 2.48 tons ha^{-1}. Over 30 tons ha^{-1} of cover crop biomass was recorded from the trials. Overall, good cover crop biomass production and weed control were observed, with no adverse effects on tomato yield, demonstrating that such a living mulch-reduced rate herbicide system is feasible.

WEED ABUNDANCE AND COMMUNITY COMPOSITION IN A LONG-TERM ORGANIC CROPPING SYSTEMS TRIAL. A.B. Jernigan* and M.R. Ryan, Cornell University, Ithaca, NY (67)

ABSTRACT

Weed management is a major constraint in organic cropping systems and research is needed to evaluate the long-term impacts of different organic management systems on weed population and community dynamics. In 2004, the Cornell Organic Cropping Systems Experiment was established in Freeville, New York using a split-plot randomized complete block design with two crop rotation entry points (split-plot factor). The experiment compared four organic vegetable cropping systems that varied in inputs and weed management strategies (main plot factor): 1) High intensity (goal was to maximize crop production with inputs), 2) Intermediate intensity (goal was to use legumes to provide nitrogen), 3) Bio-extensive (goal was to use cover cropping and fallow periods to reduce weed seedbank), 4) Reduced-tillage (goals was to use ridge tillage and controlled traffic). The High intensity and the Intermediate intensity systems simulated land limited farms that needed to invest more inputs into the soil using different strategies, the Bio-extensive system simulated a farm with land abundance that was not dependent on cropping all available land every season, and the Reduced-tillage system simulated a farm with permanent ridges that improved soil quality by not disturbing the ridge bases. All systems followed a 4-year crop rotation of winter squash, cabbage, lettuce, and potato. In the last year of the experiment a uniformity trial was conducted in which oats were grown in a uniform manner over all the system plots. Prior to sowing oats, soil samples were collected and a germination bioassay was conducted to determine if the systems differed in terms of soil weed seedbank density and community composition. After the oats reached maturity, oat biomass and weed biomass samples were collected using 0.25 m^2 quadrats. Soil weed seedbank density was affected by both cropping systems and rotation entry point and ranged from 173 (Reduced-tillage, entry point 2) to 16 (Bio-Extensive, entry point 1) seeds kg^{-1} soil. The Bio-extensive cropping system also tended to have lower weed biomass and weed density in oats when compared to the other three systems; however, the cropping systems did not differ in terms of oat biomass production. Differences in weed abundance between the four cropping systems indicate that some management practices are more effective at weed suppression than others, and that maintaining a low level of weeds in organic vegetable production is possible if weed management is prioritized.

THIN PASPALUM CONTROL IN COOL-SEASON TURF. J.R. Brewer*, S.D. Askew, and S.S. Rana, Virginia Tech, Blacksburg, VA (68)

ABSTRACT

Paspalum grass species such as thin paspalum (*Paspalum setaceum*) and dallisgrass (*Paspalum dilatatum*) plague golf courses all over the southeastern and northeastern regions of the United States. These grass species are clump-forming perennial grasses that can be tough to control in a turfgrass setting due to a system of underground rhizomes. Dallisgrass was controlled for years by MSMA primarily in warm-season turfgrass, but control often required several high-rate applications at 7 to 10 day intervals. MSMA use has recently been severely limited on golf courses due to Environmental Protection Agency (EPA) regulations. Some herbicides have shown promise for *Paspalum* spp. control in cool-season turfgrass including fluazifop, mesotrione, and topramezone. Our studies were initiated on June 18, 2015 and July 28, 2015 on a tall fescue rough at the Pete Dye River Course at Virginia Tech near Radford, VA. Both studies were formatted in a randomized complete block. These studies were established to search for effective control options for thin paspalum in cool-season turfgrass. Treatments for these studies included: topramezone at 24.6 g ai ha^{-1}, mesotrione at 175 g ai ha^{-1}, fluazifop at 105 g ai ha^{-1}, topramezone + mesotrione, and topramezone + fluazifop. All treatments were applied twice at 3 week intervals, and included methylated seed oil at 0.5 percent V/V. These treatments were applied with a hooded sprayer at 280 L ha^{-1} and 4.8 km ha^{-1}, and the sprayer had a 71 cm spray width.

Initial thin paspalum cover was 63 to 77 percent at the first site and 22 to 47 percent at the second site. At 3 WAIT, all topramezone and fluazifop treatments controlled thin paspalum greater than 88 percent at the first site and between 63 and 85 percent at the second site. Treatments were initiated later in the season at the second site and thin paspalum was more mature at initial treatment, and possibly more resistant to herbicides. Mesotrione in both studies controlled thin paspalum less than 50 percent. Tall fescue was only injured by fluazifop (13 percent) during the second study at the 3 week rating. During the 6 week rating of the first study and the 7 week rating of the second study, all topramezone and fluazifop treatments controlled thin paspalum at least 99 percent, while mesotrione controlled thin paspalum 72 and 87 percent. Fluazifop was the only treatment that caused unacceptable injury during the 6 and 7 week rating, which ranged from 43 to 57 percent.

WEED SUPPRESSION IN GRASS-LEGUME SUMMER ANNUAL FORAGE INTERCROPS. K.A. Bybee-Finley* and M.R. Ryan, Cornell University, Ithaca, NY (69)

ABSTRACT

A warm-season annual intercropping experiment was conducted across 3 locations in the Northeastern United States in 2013 and 2014 with forage species selected based on differences in stature and nitrogen acquisition traits: 1) pearl millet (*Pennisetum glaucum* L.); 2) sorghum sudangrass (*Sorghum bicolor* L. Moench x *S. sudanense* P.); 3) cowpea (*Vigna unguiculata* L.); 4) sunn hemp (*Crotolaria juncea* L.). Crops were seeded in monoculture based on recommended seeding rates and in 3 and 4 species mixtures using a replacement design where monoculture seeding rates were divided by the number of crops in the mixture. Crop and weed biomass were sampled twice by clipping vegetation in 0.5 m^2 area at approximately 45 and 90 days after planting. In general, crop biomass production in the legume monoculture treatments was lower than in the other treatments at both sampling dates, whereas biomass production in the grass monoculture treatments did not differ from the intercrop treatments. When averaged across the site-years, biomass of the monoculture treatments ranged from 1,080 and 2,700 kg ha^{-1} (cowpea) to 3,000 (pearl millet) and 8,900 kg ha^{-1} (sorghum sudangrass) for the first and second sampling dates, respectively. Weed biomass tended to decrease with increasing crop biomass and was negatively correlated with species richness, which supports agroecological theory that links weed suppression to crop diversity.

NITROUS OXIDE EMISSIONS IN LONG TERM CROPPING SYSTEMS. A.M. Knight*, W.J. Everman, S. Reberg-Horton, S. Hu, D.L. Jordan, and N. Creamer, North Carolina State University, Raleigh, NC (70)

ABSTRACT

Agriculture accounts for a large portion of land use worldwide. In the U.S. specifically, the World Bank indicated that agriculture accounts for roughly 45 percent of land use. Agriculture is estimated to contribute greatly to the output of one of the main greenhouse gases, nitrous oxide, which is suspected of contributing to climate change, contributing an estimated 59 percent to emissions. These large percentages are suspected to partially be due to one-third of nitrogen applied to cropping systems being utilized by the system while the additional two-thirds are lost to the environment. With different agricultural practices contributing to these greenhouse gas emissions, finding how various production practices contribute to greenhouse gas emissions will help in the recommendation of best management practices to minimize gas emissions by agriculture in the southeastern U.S. Field studies were conducted in 2013, 2014, and 2015 at the Center for Agricultural Farming Systems at the Cherry Research Farm in Goldsboro, NC. Long-term plots of conventional no-till, conventional-tillage, conventional crop-hay, organic tillage, organic minimal tillage, and organic crop-hay systems were used to measure the flux of the greenhouse gases CO_2, CH_4, and N_2O, 24 to 48 hours after ~1.25 cm or more of rainfall, following USDA-ARS GRACEnet Project Protocols. Preliminary results indicated that tillage and fertilization play a significant role in gas emissions in cropping systems.

EB-ESA PHD STUDENT CONTEST - ORAL PRESENTATIONS

A GENOME WIDE ASSOCIATION STUDY OF RESISTANCE TO ENTOMOPATHOGENIC FUNGI. J. Wang*, H. Lu, and R. St. Leger, University of Maryland, College Park, MD (71)

ABSTRACT

In a sister study to this one we screened more than 2,500 mutations for their effects on disease resistance. A complementary approach to mutagenesis is to identify loci at which alleles with more subtle effects segregate in natural populations. Single nucleotide polymorphisms (SNPs), insertions, and deletions in a natural population of flies are mutations that have survived the filter of natural selection and can be tested via genome-wide association (GWA) for effects on genetic variation in resistance. Here we used 203 fly lines from the Drosophila Genetic reference Panel (DGRP) to identify

SNPs associated with natural variation to disease resistance. Using topical infections with *Metarhzium anisopliae* (Ma549), we calculated LT50's, fungal loads, latent period (interval between infection and sporulation), and sporulation capacity. All parameters were genetically variable among the DGRP lines and LT50s were sexually dimorphic. We identified many SNPs in novel loci that are potentially associated with natural variation in disease resistance, as well as SNPs within genes previously known to affect resistance. Many of these loci are known to interact physically and/or genetically, enabling us to place them in candidate genetic networks. Several of the candidate genes have human homologues that were identified in studies of human disease, suggesting that genes affecting variation in susceptibility are conserved across species.

UTILIZING HERBIVORE INDUCED VOLATILES FOR ENHANCED BIOLOGICAL CONTROL: INVESTIGATING HOW THIS IDEA TRANSLATES SOIL INTERACTIONS.

M.J. Rivera*, H. Alborn, and A.M. Koppenhofer, Rutgers University, New Brunswick, NJ (72)

ABSTRACT

Recent studies have shown the enhancement of biological control when using commercially available lures that emit a herbivore-induced plant volatile. Thus far, despite advances in the understanding of their attractiveness to entomopathogenic nematodes (EPN), there has been little investigation of potential to utilize this concept for enhanced control of belowground insect pests. We compared (E)- Î²-caryophyllene (maize) and pregeijerene (citrus) in the highbush blueberry (*Vaccinium corymbosum*) system in their ability to enhance the attractiveness and enhanced efficacy of EPN against the system's herbivore, oriental beetle (*Anomala orientalis*). Using an endemic strain of the EPN, *Steinernema glaseri,* in a six-arm olfactometer, the relative attractiveness of (E)-Î²-caryophyllene (EÎ²C) and pregeijerene (PG) was tested in the lab to gather baseline values of attraction to the chemicals alone in sand substrate. One week before the field study, the soil of 30 plants was sampled and baited with *G. mellonella* larvae to select plants with the highest *S. glaseri* activity. An arrangement similar to the 6-arm olfactometer was used in the field by placing six gasket traps containing third instar oriental beetle larvae or G. mellonella larvae with and without compound into the soil around the base of 10 plants. The gaskets were removed after 72 hours and insect baits retrieved and assessed for EPN infection. After removal of the insect bait from the gasket, the remaining clean sand packed in the gasket along with the insect was exhaustively baited with *G. mellonella* larvae to assess EPN density. The lab results suggest that in sand alone EÎ²C is significantly more attractive than PG to endemic *S. glaseri*. In the field, there was no difference in attractiveness or efficacy in the field study but rather, endemic *S. glaseri* were more attracted to gasket traps without oriental beetle larvae than traps with oriental beetle larvae. Similarly, when this field study was replicated with a widely susceptible insect host, *Galleria mellonella*, more EPN successfully infected the control than either treatment.

INTERSPECIFIC LARVAL INTERACTIONS BETWEEN *DROSOPHILA SUZUKII* AND *ZAPRIONUS INDIANUS* IN FOUR VARIETIES OF VIRGINIA WINE GRAPES. M.E. Shrader* and D. Pfeiffer, Virginia Tech, Blacksburg, VA (73)

ABSTRACT

The insect pest ecology within Virginia vineyards has changed dramatically over the past decade with the introduction of several new invasive species. The latest introductions have been the spotted wing drosophila (SWD), *Drosophila suzukii* and the African fig fly (AFF), *Zaprionus indianus*. SWD is capable of ovipositing into intact immature grapes, while AFF needs a cracked or wounded grape in which to deposit its egg. Since SWD is capable of wounding the grape it is possible for AFF to use these oviposition wounds to deposit its eggs onto the surface of the grape. Thus, it is possible to have both fly species developing within a single grape berry. This dual infestation compelled research to determine the impact of AFF interactions on the development of SWD in Virginia wine grapes. Six grape varieties were selected from a single vineyard in the Piedmont region of Virginia. The larval interactions were assessed using four larval densities consisting of 1:1 and 2:2 (SWD : AFF) and controls were tested at 2 and 4 SWD per grape. This research evaluated the larval interactions of these two species within a single grape by determining total development time, pupal volume, larval mortality and pupal mortality. Results from this study indicated that interspecific larval competition increased development time and decreased pupal volume of SWD within wine grapes compared to the controls. The presences of AFF larvae also increased both larval and pupal mortality of SWD. Wine grape varietal differences were also noted when assessing these larval interactions.

DIFFERENTIAL SUSCEPTIBILITY OF FALL ARMYWORM (*SPODOPTERA FRUGIPERDA* SMITH) HOST-STRAINS TO SELECTED BT TOXINS. D.A. Ingber* and C.E. Mason, University of Delaware, Newark, DE (74)

ABSTRACT

The fall armyworm (*Spodoptera frugiperda* Smith) is a polyphagous, multivoltine pest of several commercial crops including corn, cotton, sorghum, rice, and bermuda grass. Populations overwinter in the tropical regions of South and Central America, Mexico, and the Southern United States and migrate to more northern latitudes yearly. Fall armyworm is of particular concern in insect pest management as populations have developed resistance to the Cry1F toxin derived from the bacterium *Bacillus thuringiensis* (Bt) produced by transgenic corn (*Zea mays* L). Additionally, fall armyworm consists of two host-strains that are referred to as the corn and rice strains due to host preferences and the plants that they were originally collected from. The strains differ in wing morphometrics, but are more commonly distinguished using genetic markers. There is some evidence that the corn and rice strains differ in their tolerances to select insecticides; this extends to several Bt toxins, including Cry1Ab and Cry1Ac, though results are mixed. In this study, a series of diet-based bioassays were conducted with corn and rice strains of fall armyworm to test for differential susceptibility to Cry1Ab, Cry1Ac, and Cry1F. Mortality and growth inhibition data were collected, and used to calculate LC_{50} and EC_{50} values. Data on differential fitness between the host-strains of fall armyworm could aid in the development of regional pest and resistance management strategies in locations where each host-strain is known.

INJURY TO APPLES AND PEACHES AT HARVEST FROM FEEDING BY *HALYOMORPHA HALYS* (HEMIPTERA: PENTATOMIDAE) NYMPHS EARLY AND LATE IN THE SEASON. A.L. Acebes-Doria*, T.C. Leskey, and J.C. Bergh, Virginia Tech AREC, Winchester, VA (76)

ABSTRACT

Adults and nymphs of *Halyomorpha halys*, feed on tree fruits. Feeding injury from adults has been characterized, but the injury from nymphs has not been examined systematically. We compared feeding injury at harvest from 2nd and 3rd instar nymphs, 4th and 5th instar nymphs, and adults that were caged on 'Smoothee Golden' apples and 'Redhaven' peaches in early June (peach and apple), late July (peach), and late August (apple). Individual apples and peaches were caged at fruit set and assigned to the following treatments (n = 28 fruit/treatment): 1) control (no bugs), 2) early-season young nymphs and 3) adults, and 4) late-season young nymphs, 5) older nymphs and 6) adults. Fruit in each treatment were exposed to 3-4 young nymphs, 2-3 older nymphs or 1-2 adults for 96 h and and evaluated for external and internal feeding injury within 36 h after harvest. For both apples and peaches, no injuries were recorded from control fruit, but the percentage of injured fruit and number of injuries per fruit varied across all exposed treatments. Among those, early-season feeding of young nymphs yielded the least injury. In apples, the highest percentage of injured fruit and number of injuries per fruit were caused by late-season feeding by adults. In peaches, early-season adult feeding produced the highest percentage of injured fruit and injuries per fruit. Our results revealed new information that has implications for *H. halys* management (*e.g.* timing of intervention) in fruit orchards.

WHAT DO BIRDS EAT? INVESTIGATING INSECT PREY PREFERENCES AND THEIR CONSEQUENCES IN NORTH AMERICAN SONGBIRDS. A.C. Kennedy* and D. Tallamy, University of Delaware, Newark, DE (77)

ABSTRACT

Ninety-six percent of North American terrestrial bird species rear their young partially or exclusively on insects, yet we know little about which insects they prefer or how plastic their diets are. Most ornithological references provide only generalized summaries of which arthropods are consumed by birds (e.g., "caterpillars", "beetles"), rather than species-specific or even family-level identification of prey. Without such knowledge, experts are limited in their ability to manage landscapes effectively for avian conservation. To restore viable bird habitat, we need to know which insect species birds eat, and how plastic their diets are, so that we can establish the host plant species of those insects. To address this, nest cameras were placed at Eastern Bluebird and

House Wren nests at Mount Cuba Nature Center (Hockessin, DE) to record insect prey delivered to nestlings. Additionally, a citizen science project was launched to crowd-source photos of birds taking insects to their nests. Preliminary results suggest that songbirds rely heavily on Lepidoptera in their diets, perhaps as a critical source of carotenoids. Ultimately, this project will yield recommendations for which host plant species will promote the growth and survival of select songbird species.

DIFFERENCES IN FITNESS TRAITS BETWEEN E AND Z PHEROMONE RACES OF EUROPEAN CORN BORER, *OSTRINIA NUBILALIS*, ON DIFFERENT HOST PLANTS.
H.L. Walker* and C.E. Mason, University of Delaware, Newark, DE (78)

ABSTRACT

Life-history traits can have a direct impact on an organism's overall fitness. By looking at differences in life-history traits of closely related insect groups we can better understand what drives these changes and how they affect fitness. The European corn borer (ECB), *Ostrinia nubilalis* (Hübner), is a polyphagous insect species native to Europe, where it is widespread. ECB is considered a major economic pest of corn in the United States and Canada. ECBs have two distinct pheromone races, the E and Z races, which produce different ratios of two isomers in their pheromone blends. The pheromone races also can exhibit variation in emergence behavior and host plant selection. This research focuses on how E and Z ECB larvae differ in fitness due to changes in life-history traits on a range of host plant based diets. Assessment includes ECB larvae from both laboratory colony and wild captured populations. Identification of pheromone race of ECB individuals is done using a PCR and restriction enzyme method. ECB larvae are reared on either a Cry1F Bt corn, non-Bt corn, sweet corn, tomato, or hops variety plant-incorporated diet. Larvae are then measured for variations in: (1) growth and development, (2) pheromone production, (3) fecundity, and (4) survival. Data are analyzed using two-way ANOVA to determine if there are significant differences in life-history traits between the pheromone races.

LIFE STAGE SPECIFIC PREDATION OF *HALYOMORPHA HALYS* (STAL). J.M. Pote* and A.L. Nielsen, Rutgers University, New Brunswick, NJ (79)

ABSTRACT

Halyomorpha halys is a major pest of American fruit, vegetable and ornamental crops. Due to its history as an invasive species, little is known about the predators affecting *H. halys* in the US. We conducted no-choice predation experiments on several species of predators and omnivores commonly found in crops affected by *H. halys*. Our results indicate that many predators of *H. halys* confine their consumption of this pest to a single life stage. Additionally, we surveyed three crops throughout the 2015 growing season for predator abundance. By incorporating preference data from the lab and abundance data from the field, we are able to hypothesize which predator species will be temporally synchronized with prefered/susceptible *H. halys* life stages.

SOIL PROPERTY EFFECTS ON ARBUSCULAR MYCORRHIZAL FUNGI (AMF) AND DARK SEPTATE ENDOPHYTE (DSE) COLONIZATION OF *PANICUM VIRGATUM* L. (SWITCHGRASS). M.R. Bindell* and N. Zhang, Rutgers University, New Brunswick, NJ (80)

ABSTRACT

Arbuscular mycorrhizal fungi (AMF) and dark septate endophytes (DSE) often co-occur in nature; however, it is unclear what factors contribute most to their colonization of plant roots. To better understand the effects of soil properties on AMF and DSE colonization, 58 root samples of *Panicum virgatum* L. (switchgrass) were collected from natural and managed landscapes of varying soil properties (i.e. soil pH and nutrient levels) and observed microscopically. Sampling locations included the Pine Barrens of New Jersey, the Pine Barrens and Dwarf Pine Plains of Long Island, NY, an Iowa wetlands site, and two switchgrass research plots in New Jersey. We expected soil properties to impact AMF and DSE colonization. Results showed significant differences in percent AMF colonization in natural versus managed locations within NJ and also differences among acidic versus non-acidic soils. The NJ Pine Barrens contained the highest percent colonization of AMF. Cloning techniques were employed to gain further analysis of the Pine Barrens AMF. *Glomus* sp. was the most common AMF sequenced in the NJ Pine Barrens. However, there were several AMF that had <97 percent similarity to known sequences and therefore may be undescribed species. This is the first report of AMF and DSE in the LI Pine Barrens and the first detailed report of AMF in the NJ Pine Barrens. This study shows that soil properties may impact AMF and DSE abundance. Work in progress includes a metagenomic analysis of these fungal communities. Future work should include phylogenetic analysis of the aforementioned, undescribed Pine Barrens AMF.

EFFECT OF NITROGEN FERTILIZER APPLICATION RATE ON YIELD AND QUALITY OF BRACHYTIC DWARF BROWN MIDRIB (BMR) FORAGE SORGHUM . S.E. Lyons*, T. Kilcer, G.S. Godwin, D. Cherney, J. Cherney, and Q.M. Ketterings, Cornell University, Ithaca, NY (81)

ABSTRACT

Winter cereals grown as forage double crops in corn silage rotations in New York State during the past several years have yielded, on average, 3.6 Mg dry matter (DM)/ha for cereal rye and 4.9 Mg DM/ha of triticale. However, incorporation of these winter forages into corn silage rotations can be a challenge in the relatively short growing seasons of the Northeastern United States. Brachytic dwarf brown midrib (BMR) forage sorghum was investigated as an alternative to corn silage due to its high yield potential and relatively short growing season, allowing for the inclusion of winter cereals in rotation without affecting the yield of the main crop. Forage sorghum was grown at 5 N rates (0, 56, 112, 168, and 224 kg N/ha, 4 replicates each) during the summers of 2012-2015 in eastern and central NY (9 locations). Plants were harvested at soft-dough stage and analyzed for DM yield and crude protein (CP). In addition, neutral detergent fiber (NDF), total digestible nutrients (TDN), and starch content were measured for three of the sites. Up to 17.5 Mg DM/ha was achieved at the optimum economic N rate ($P = 0.0001$) but yields varied from site to site and optimum N rates ranged from 0 to 168 kg N/ha depending on the location. The yields of two sites were unresponsive to N treatment ($P > 0.3$), while for the other seven locations, a crop yield response to N addition was measured ($P < 0.05$). Crude protein increased with increasing N rates ($P < 0.0001$), reflecting additional N uptake by the plants ($P < 0.0001$). There was no difference in NDF, TDN, or starch content among N treatments ($P > 0.1$). Results to date suggest that brachytic dwarf brown midrib forage sorghum is a viable alternative to corn silage for summer forage production.

DUAL PURPOSE COVER CROPS FOR YIELD INCREASES AND MANURE NUTRIENT MANAGEMENT ON NEW ENGLAND DAIRY FARMS. S. Corcoran*, University of Massachusetts Amherst, Amherst, MA (82)

ABSTRACT

Cover crop benefits are well known, but many New England dairy farmers do not prioritize planting cover crops. Cover crops are planted following full season corn, which is too late to offer significant benefits, or cover crops are not planted at all. However, forage production is the largest annual expense on Northeast dairy farms. Therefore, dual-purpose cover crops harvested for additional forage in the spring provides incentive to plant fall cover crops early following shorter-season corn. To assess the dual-purpose cover crop strategy, two field experiments were conducted in fall 2014-summer 2015 following fall manure application.

Experiment 1: Rye was planted on 9/1, 9/15, and 9/30 followed by shorter, mid, and full season corn to assess the yield potential spectrum of all combinations.

Experiment 2: Rye, wheat, and triticale were planted on 9/1, followed by early spring N applications to assess crop response. Shorter-season corn was subsequently planted to assess each crop's impact on corn production for silage.

These ongoing experiments study: 1. manure nutrient management with specific interest in N&P, 2. the effect of planting date on nutrient capture and fall and spring yield, 3. a direct comparison of rye, wheat, and triticale performance, 4. impact on corn production, 5. economic value of the dual-purpose crop, 6. whole system yield and management.

Our preliminary results suggest farmer profitability or savings potential over \$500/acre , improved manure management and nutrient capture, alternative forage sources offering resiliency to climate change, and better management practices leading to increased sustainability and environmental benefits.

USING ELECTRICAL RESISTIVITY IMAGING TO CHARACTERIZE SUBSURFACE PHOSPHORUS MOVEMENT TO DRAINAGE DITCHES. K.L. Clark*, A. Shober, A. Buda, J. Robinson, and S. Andres, University of Delaware, Newark, DE (83)

ABSTRACT

Long-term application of poultry litter to agricultural soils on Delmarva has led to the build-up of soil test phosphorus (P) and increased risk of P losses to sensitive water bodies. Subsurface lateral flow pathways can deliver significant P loads from high P agricultural soils with artificial drainage. However, the mechanisms of subsurface P transport are poorly understood. For this study, salt tracers were paired with electrical resistivity imaging (ERI) to determine the site characteristics and environmental conditions that affect subsurface lateral flow of water to drainage ditches. Initially, soil cores and water samples collected from an agricultural field with a history of poultry litter application at the University of Maryland Eastern Shore research farm were analyzed for water soluble P and total dissolved P. Within the same field, a potassium bromide tracer was introduced 10.5 m from a drainage ditch via a 25-cm deep trench. Movement of the tracer toward the ditch was monitored from May 2015 through December 2015 using ERI. Electrical conductivity sensors were placed in 1 m deep wells along the suspected flow path in order to confirm ERI-observed salt plume movement. Very little movement of the bromide tracer occurred during the initial observation period, suggesting that nutrient movement was minimal when the water table was deep. Continued observations should show more rapid movement of the tracer during wetter periods with higher water tables. Results of this study are expected to provide a better understanding of P movement in artificially drained systems to guide agricultural P management.

MAXIMIZING THE ON-FARM BENEFITS OF COVER CROPS: STUDYING THE BEST MATCHES BETWEEN MANAGEMENT INTENTIONS AND ECOSYSTEM SERVICE PROVISIONING. A.V. Hamilton* and D.A. Mortensen, Penn State University, State College, PA (84)

ABSTRACT

Cover crops are gaining in popularity due to efforts from both conservation organizations and farmers for the multiple ecosystem services they can improve. Much is known with regards to ecosystem service provisioning from cover crops, but the information comes mainly from highly controlled field experiments. Published information on farmers' cover cropping practices and perspectives are almost exclusively conducted with midwestern farmers. The purposes of this study were to: 1) Study farmers' cover cropping practices on-farm to determine the extent to which they influence their cover crops' provisioning of ecosystem services, and 2) Characterize east coast attitudes and practices with regards to cover cropping.

Qualitative and quantitative data was collected from 47 diverse farms in Pennsylvania between Fall 2014 and Spring 2015 via a farmer questionnaire and on-farm cover crop sampling. Farmer participants selected several of their farm's fields of cover crops for analysis, and recorded their intentions for planting those cover crops upon planting, and their perceptions of the field's contributions upon spring termination. The species, biomass and nutrient content of the cover crop at time of termination were used to calculate the field's potential contributions to five ecosystem services: Organic matter provision, erosion reduction, weed suppression, nitrogen provision and nitrogen scavenging. Each field's intention, perception and ecosystem service assessment ratings were compared and demonstrated no relation between a fields' level of ecosystem service provisioning and the farmer's intention or perception of those services. Nitrogen scavenging was the only ecosystem service that demonstrated a significant relation between farmers' intention rating and the field's extent of provisioning the service. Forty to 60 percent of the fields did not achieve the farmer's desired level of ecosystem service provisioning for any given ecosystem service, although the majority of fields provided a moderate to high level of erosion control, organic matter provisioning and nitrogen scavenging.

EFFECT OF FERTILIZER TIMING AND RATE ON THE YIELD AND NUTRIENT USE EFFICIENCY OF IRRIGATED CORN IN DELAWARE. A. Soroka*, University of Delaware, Newark, DE (85)

ABSTRACT

Inefficient use of nitrogen (N) in agronomic crop production can lead to water quality concerns and reduced yields for growers. Nitrogen fertilizer left in the soil after harvest is highly mobile and can leach to groundwater or is subject to gaseous loss. Split applications of N can better match N fertilizer applications with corn N uptake needs and improve N use efficiency. The objective of this study was to quantify the effect of N rate, timing, and application method under central pivot and subsurface drip irrigation on corn grain yield and N use efficiency in Delaware. All plots (with exception of the 0 N control) received pre-plant manure (94 kg ha^{-1} plant available N) and starter N fertilizer (N rate = 34 kg ha^{-1}). Additional plots received in-season N fertilizer applications below (185 kg ha^{-1}), at (240 kg ha^{-1}), or above (290 kg ha^{-1}) current Delaware N recommendations for a realistic goal of 16 Mg ha^{-1}. Grain yield was determined for each treatment using a weigh wagon. Pre-plant soil, post-harvest grain and residue were collected then analyzed for total N to determine N use efficiency by difference and mass balance methods. In 2014, plots receiving in-season applications of N out yielded plots receiving only manure and starter N (17 *vs.* 15 Mg ha^{-1}). Timing and rate of N application in-season did not affect yield. Exceptional growing conditions pushed N efficiency values over 100 percent for most treatments, suggesting mineralization of manure N was adequate to sustain high yields.

IMPACT OF FRAZE MOWING ON BERMUDAGRASS RECOVERY FROM SPRING DEAD SPOT. C.D. Shelton*, D.S. McCall, and G.L. Miller, Virginia Tech, Blacksburg, VA (86)

ABSTRACT

Spring dead spot (SDS), caused by *Ophiosphaerella spp.*, is one of the most prominent diseases of bermudagrass. Managing this disease with fungicides can be difficult and requires sequential applications over multiple years. SDS is characterized by circular depressions of voided turf varying in size up to one meter. Under heavy pressure, the voids result in an uneven surface that decrease playability. "Fraze mowing" is an emerging strategy which removes most surface plant material and thatch allowing for a smoother playing surface. Our objective is to explore alternative options for the recovery and long-term suppression of spring dead spot with fraze mowing. Research trials were conducted at the Strikers Soccer Club West Creek Fields in Richmond, VA in 2015. Plots were 4.5 m x 1.8 m and arranged in a randomized complete block design with four replications. Fraze mowing was conducted on 5/29/2015 at four and eight mm depths using a Koro Fieldtopmaker. Urea and ammonium sulfate were applied as granular broadcast applications immediately following trial initiation every seven days for six weeks at a rate of 24.4 kg N ha^{-1}. Visual assessments and reflectance data were collected during the six week application period. The simple ratio vegetation index (RVI= NIR 760nm/ RED 670nm) was calculated with reflectance data and transformed to show objective changes in plot quality over time. Area under turf quality progression curve was calculated for all treatments using RVI data. RVI values were lower in fraze mowed plots for the first two weeks after application. Fraze mow treatments yielded higher turf quality (reflectance and visual) on subsequent dates. Six weeks after fraze mowing, plots treated with ammonium sulfate recovered faster than plots treated with urea, regardless of depth. Spring data collection will be used to determine whether fraze mowing impacts the SDS epidemic after a growing season.

HOST ADAPTATION AND MOLECULAR DETECTION OF *SCLEROTINIA HOMOEOCARPA*. B.A. Aynardi*, M.M. Jimenez-Gasco, and W. Uddin, Penn State University, University Park, PA (87)

ABSTRACT

Dollar spot of turfgrass, caused by *Sclerotinia homoeocarpa* F.T. Bennett, is one of the most common and most costly diseases of high input turfgrasses, particularly on golf courses. A molecular detection method for *S. homoeocarpa* has been developed through the synthesis of species-specific primers for use in conventional and quantitative PCR. Additionally, Internal transcribed spacer (ITS) region sequence analysis has identified two genetically distinct types of isolates; a Type I group, which consists predominately of C3 isolates (collected from cool-season turfgrass hosts), and a Type II group, comprised of only C4 isolates (collected from warm-season turfgrass hosts). Both group type isolates have been found to coexist in the same locale, and tend to infect the preferred host type. However, in northern areas, C4 isolate analysis indicates that these isolates group genetically with Type I isolates. Therefore, the purpose of this presentation is to determine the extent at which *S. homoeocarpa* isolates infect and cause disease on cool and warm-season grasses that co-inhabit specific locations, the extent to which warm-season isolates have moved north, and the ability of these isolates to cause disease on other host types under specific environmental conditions. Preliminary results indicate that C3 isolates cause higher disease severity on both host types under varying environmental conditions. This is of importance as warm-season species, such as bermudagrass, are being used more widely in northern locations.

TRENDS IN SOIL TEST PHOSPHORUS DYNAMICS FOLLOWING LONG-TERM APPLICATION OF POULTRY LITTER AND COMMERCIAL FERTILIZERS. Z. Qin*, A. Shober, and S. Tingle, University of Delaware, Newark, DE (1)

ABSTRACT

Historical application of poultry litter to Delmarva soils resulted in build-up of soil test phosphorus (P) above agronomic optimum levels and increased risk of P loss. We aim to understand P dynamics during build-up of soil test P in agricultural soils receiving long-term applications of manure or inorganic P fertilizers. Selected soil samples (2000-2013) were collected from plots at two field sites receiving no P, poultry litter at moderate and high rates (224 and 897 g m^{-2}), or triple superphosphate (TSP; annual P_2O_5 rate = 6.73 and 13.5 g m^{-2}) for 15 years. Soils were analyzed for water extractable P (WEP), Mehlich-3 P (M3P), and Mehlich 3 P saturation ratio (PSR); a sequential P fractionation (H_2O, $NaHCO_3$, NaOH, and HCl) was also completed. Soils receiving the high manure rate at both sites exhibited flat or increased WEP and M3P concentrations, while these concentrations often decreased in soils receiving lower manure or TSP rates. In most cases, soils receiving the high manure and/or TSP rates had significantly higher WEP (4.42 mg kg^{-1}), M3P (123 mg kg^{-1}), and PSR (28.2 percent) than soils receiving moderate P rates or no P (averaged 0.86 and 71.5 mg kg^{-1}, and 18.5 percent for WEP, M3P, and PSR respectively). Repeated application of manure and TSP at rates exceeding crop removal (high rate) increased P concentrations in both the labile (H_2O-P and $NaHCO_3$-P) and recalcitrant soil P pools (HCl-P). Diligent management of legacy P soils is necessary to control potential dissolved P release.

NEWSS STUDENT CONTEST - ORAL PRESENTATIONS (II)

A RETURN TO NORMAL: LEGACY EFFECTS OF JAPANESE STILTGRASS. D.R. Tekiela* and J.N. Barney, Virginia Tech, Blacksburg, VA (89)

ABSTRACT

Invasive plants have been shown to negatively impact most facets of ecosystem function, services, and health, and cost the United States millions of dollars annually. As a result of these impacts, management and control of invasive plants has been highly prioritized in recent years. Although many resources are spent attempting to control invasive plants, little is known about the effectiveness of managing an invasive plant to return a system to its uninvaded state. Also, little is known about the effect of duration of invasion on the level of impact invasive plants impose. Therefore, we examined the effect of managing an invasive grass, Japanese stiltgrass (*Microstegium vimineum*), on the return of the plant community to its uninvaded state and the effect of establishing a new Japanese stiltgrass population in an uninvaded landscape. In other words, does the plant community return to an uninvaded state following removal of the invasive plant.

Using an observational and removal method in three locations across the Virginia Blue Ridge Mountains, we found that newly established Japanese stiltgrass invasions had little effect on the resident plant community in three years and remained similar to uninvaded plots suggesting a potential temporal threshold before Japanese stiltgrass imposes resident plant community impacts. However, following Japanese stiltgrass removal, the plant community did not return to an uninvaded state. In fact, the resident plant community in invaded plots was more similar to uninvaded plots than removal plots were to uninvaded plots after three years of invader removal. Invader removal plots also had greater overall variation in plant community composition than any other treatment suggesting instability in the plant community. All effects of Japanese stiltgrass removal occurred within the first year; however this new plant community composition remained stable for three years. Within a three-year period, removal of Japanese stiltgrass does not restore the plant community and instead creates a novel plant community.

WEED AND CROP RESPONSE TO INTEGRATED MANAGEMENT IN A DIVERSE CROPPING SYSTEM. K.E. Caswell*, W.S. Curran, and H. Karsten, Penn State, University Park, PA (90)

ABSTRACT

A diverse, no-till dairy cropping systems study was established in 2010 seeking to produce enough feed, forage, and fuel to supply a 65-cow, 97 hectare dairy farm in Pennsylvania while minimizing off farm inputs. One of the 6-year diverse crop rotations, the Pest Management rotation, evaluated strategies that attempted to reduce herbicide inputs and the risk of herbicide resistant weeds. A Reduced Herbicide (RH) management was compared to a Standard Herbicide (SH) management that as more reliant on chemical weed control. Both managements utilized Integrated Weed Management with the RH management incorporating more cultural and mechanical controls than the chemical based SH management. The crop rotation was composed of annual grain crops, an oil seed crop, cover crops and perennial forages. In 2015, the full six year rotation was completed, allowing for the comparison of the response to the both the rotation and different managements. The RH and SH managements were compared based on crop yield and weed biomass. These performance indicators were reflective of management changes made in 2013 in efforts to maintain or improve weed control and crop establishment. Weed control in SH management was excellent and end of season weed biomass never exceeded 10 kg/ha in either corn or soybean. End of season weed biomass in the RH management was either equal or greater than SH; weed biomass exceeded 300 kg/ha in two of six years in soybean and averaged 150 kg/ha in corn. Crop yield varied between the rotations in several years, but these differences were generally not attributed to weed competition. In conclusion, the RH treatment was not as effective at controlling weeds, but maintained weed biomass below crop limiting levels, while reducing herbicide inputs.

AUTUMN OLIVE GERMINATION: AN UNLIKELY STORY. M.E. Franke* and J.N. Barney, Virginia Tech, Blacksburg, VA (91)

ABSTRACT

Autumn olive (*Elaeagnus umbellata*) is a prolific invasive plant species in the Eastern United States, and is one of the most common invaders in the Appalachian region. It is a large shrub native to Eastern Asia that can produce hundreds of thousands of drupes per mature plant. We collected drupes and followed a cold stratification protocol to germinate the seeds in the fall of 2014, but had a germination rate <5 percent. Since the drupes are known to be a popular food for wildlife, and persist on the branches until winter, providing nourishment when other food sources are scarce (Darlington 1994; Lemke et al., 2013), we decided to further investigate the role animal species could play in the dispersal and germination of autumn olive seeds. Autumn olive is often found clumped along roads under power lines and fence lines, suggesting that birds perch on these structures are defecating viable seeds. Recognizing this pattern anecdotally, we decided to conduct experiments simulating a bird gut using sulphuric acid to mimic a bird's stomach acid and an orbiter cycler with rocks to simulate a bird's gizzard. We created a full factorial experiment in which each treatment included varying amounts of time in acid (0, 5, 30, 60, 120, or 240 minutes) or undergoing scarification (0, 2, 12, or 24 hours). We conducted this experiment in sterile conditions and planted the seeds on plates of 0.25x Hoagland solution, as well as replicating the entire experiment in a greenhouse, planting the seeds in a sandy soil mix. To ensure our autumn olive seeds were viable, we conducted a Tetrazolium Chloride (TZ) test and found 66 percent viability. Our preliminary results suggest that the longer a seed is exposed to acid (120 or 240 minutes) or scarification (12 or 24 hours), the less likely it is to germinate compared to spending shorter amounts of time in acid (5 or 30 minutes) and time undergoing scarification (2 hours). Our preliminary results suggest bird guts enhance autumn olive germination.

NED-APS STUDENT CONTEST - ORAL PRESENTATIONS

EFFECTS OF CHEMICAL AND BIOLOGICAL PRODUCTS ON PINK ROT OF POTATO. X. Zhang*, H. Jiang, S.B. Johnson, J. Hao, N.F. Marangoni, and X. Zhang, University of Maine, Orono, ME (93)

ABSTRACT

Pink rot, caused by *Phytophthora erythroseptica*, has been a persistent problem in potato production. Controlling pink rot becomes a challenge as mefenoxam is losing its effectiveness owing to the development of resistant populations of *P. erythroseptica*. To evaluate new strategies of pink rot management, field trials were conducted in 2014 and 2015 in Maine. *Phytophthora erythroseptica* inoculum was prepared in vermiculate mixed with V8 broth at 2:1 ratio (v:v) and incubated for four weeks at 22Â°C. Potato (cv. â€˜Russet Norkotahâ€™) seed tubers were planted at one-foot spacing. Inoculum was evenly distributed in the furrows at planting. Chemical products and biological control agents were applied either individually or in combinations, at planting or during the growing season, treatment dependent. Plant vigor and emergence were evaluated twice during the season. Tuber yield and disease severity were evaluated at harvest. Mefenoxam, fluopicolide, and oxathiapiprolin significantly reduced pink rot in the harvested tubers. Biologicals including *Bacillus subtilis* (Serenade Soil, Taegro), *Bacillus amyloliquifaciens* (Double Nickel, MBI-110), and extract of *Reynoutria sachalinensis* (Regalia) did not significantly reduce pink rot in the harvested tubers. However, the combination of either Presidio or A20941 andÂ *Bacillus* sp. product significantly reduced pink rot in the harvested tubers. These results demonstrate the advantage of combining synthetic fungicides with biologicals in pink rot control. Â

SEARCH FOR MOLECULAR MARKERS IN POTATO GERMPLASM FOR PTNRD EXPRESSION. W.L. da Silva*, S. Gray, and W. De Jong, Cornell University, Ithaca, NY (94)

ABSTRACT

Potato tuber necrotic ringspot disease (PTNRD) is a tuber deformity associated with infections by the tuber necrotic strain of potato virus Y (PVY^{NTN}). It negatively impacts tuber quality, reduces trade opportunities, and poses a serious threat to potato seed and commercial production worldwide. Recent surveys indicate that PVY^{NTN} now predominates in Europe and is on the rise in the USA. Our research goal is aimed at developing reliable molecular markers that will allow breeders to rapidly and economically screen early selections to eliminate PTNRD-susceptible clones. The cultivars 'Waneta' and 'Pike' express severe and mild PTNRD symptoms, respectively, when infected with PVY^{NTN} isolate NY090029. An F1 Waneta x Pike population is being used to map the genes controlling PTNRD expression. Over 200 F1 progeny were grown in the greenhouse and inoculated with PVY^{NTN} isolate NY090029 at the 8-10 leaf stage. Foliar symptom type and severity were monitored for 10 weeks; at maturity, tubers were harvested and evaluated for PTNRD expression. The progeny segregated for type of foliar symptoms and PTNRD expression with 99 percent, 39 percent, and 14 percent of the clones expressing foliar mosaic, veinal-necrosis, and foliar-necrosis symptoms, respectively. Tubers harvested from 40 percent of the clones expressed PTNRD. DNA was extracted from each clone and genotyped at 8303 SNP loci. After filtering, 1553 high quality polymorphic SNP markers are being used for linkage mapping using TetraploidMap for Windows.

ABUNDANCE OF *SCLEROTINIA HOMOEOCARPA* IN THE THATCH AND SOIL ON ORGANIC AND CONVENTIONAL TURF IN MASSACHUSETTS. E. Allan-Perkins*, D. Manter, and G, University of Massachusetts Amherst, Amherst, MA (95)

ABSTRACT

Sclerotinia homoeocarpa causes the most economically damaging disease on cool-season turfgrasses: dollar spot. The life cycle of this pathogen is poorly understood. Past reports have hypothesized that *S. homoeocarpa* overwinters as mycelia or stromata in leaf tissue and plant debris. One recent study cultured *S. homoeocarpa* from 80-100 percent of toothpicks inserted in the thatch layer of an infection center, but not from toothpicks inserted in the soil. Therefore, we decided to compare the abundance of *S. homoeocarpa* isolated from the thatch and soil from organic and conventional golf courses. We collected samples from three management areas (putting greens, fairways, and roughs) on three golf courses (one organic and two conventional) in Massachusetts in the spring and fall of 2014 and from a fairway and a putting green at the Joseph Troll Turfgrass Research Center in Massachusetts in 2015. We used quantitative PCR to estimate the amount of detectable *S. homoeocarpa* using species-specific primers. There were no significant differences in the amount of fungus detected among management areas or golf courses. However, we found there was highly significantly more *S. homoeocarpa* detected in the thatch layer (0-21.13 pg/µl) than the soil (0-0.805 pg/µl) in all samples. This result may have important implications for future studies aimed at quantifying this pathogen and for developing new management strategies aimed at reducing *S. homoeocarpa* inoculum.

TETRACYCLINE RESISTANCE GENES IN EPIPHYTIC BACTERIA COLLECTED FROM PENNSYLVANIA STONE FRUIT ORCHARDS. S.J. Bardsley Capasso*, K. Peter, H.K. Ngugi, and M.M. Jimenez-Gasco, Penn State University, University Park, PA (96)

ABSTRACT

In 2012 and 2013, 648 isolates of epiphytic bacteria were collected from 6 conventional and 2 organic commercial stone fruit orchards in Adams, Lancaster, Chester, and Delaware Counties, Pennsylvania. These bacteria were identified to genus level using 16S ribosomal sequences and screened for the incidence of tetracycline resistance genes, *tet*(A), *tet*(B), and *tet*(C). The association of management factors, including oxytetracycline use, oxytetracycline application method (e.g., alternate row middle versus complete application), and tree age, with the incidence of tetracycline resistance genes was evaluated. A total of eight bacterial genera were identified and included *Pantoea* (39.8 percent), *Xanthomonas* (31.9 percent), *Pseudomonas* (15.7 percent), *Bacillus* (6.3 percent), *Curtobacterium* (2.9 percent), *Staphylococcus* (2.6 percent), *Frondihabitans* (0.6 percent), and *Rahnella* (0.3 percent). *Pantoea* spp. were found in all eight orchards while *Bacillus* spp. were found in only the organic orchards. Tetracycline resistance genes, *tet*A, *tet*B, and *tet*C, were found in five of the eight sampled orchards. *Tet*B was most commonly associated with *Pantoea* spp. while *tet*C was most often found in *Pseudomonas* spp. The incidence of tetracycline resistance genes significantly differed ($P > 0.0001$) among the sampled orchards. The number of oxytetracycline applications made in the year of and in the year prior to sample collection was not significantly associated with the incidence of resistance genes, *tet*A, *tet*B, and *tet*C ($P = 0.0855$); however, tree age ($P > 0.0001$) and oxytetracycline application method ($P = 0.0002$) were. A greater percentage of tetracycline resistance genes were recovered from old trees (> 10 years) compared to young trees as well as from trees that had been sprayed with the alternate row middle method.

EB-ESA BS/MS STUDENT CONTEST - ORAL PRESENTATIONS

EFFECT OF HOST PLANT RESISTANCE AND INTERCROPPING ON POTATO LEAFHOPPER (*EMPOASCA FABAE*) IN ALFALFA. E.S. Keyser*, Ursinus College, Collegevillle, PA (2)

ABSTRACT

The potato leafhopper, *Empoasca fabae*, is a key pest of alfalfa in the Northeastern United States. Insecticides are an effective management tool, but are environmentally damaging. Two alternative pest management strategies are host-plant resistance and intercropping. We examined the effectiveness of three planting treatments on potato leafhoppers: Â leafhopper-susceptible alfalfa, leafhopper-resistant alfalfa, and leafhopper-resistant alfalfa intercropped with orchardgrass (n=8 for each treatment). Potato leafhoppers were monitored by sweep sampling in 2014 and 2015. In both years, leafhopper abundance was significantly higher in leafhopper-susceptible alfalfa, but there was little difference between the leafhopper-resistant and the intercropped leafhopper-resistant treatments. Thus, host plant resistance may be a viable alternative to insecticides for potato leafhopper control, but an orchardgrass intercrop appears to provide little additional benefit when using leafhopper-resistant alfalfa.

PEST MANAGEMENT OF AN INVASIVE MANGO-FEEDING FRUIT FLY, *BACTROCERA DORSALIS* (DIPTERA: TEPHRITIDAE), IN SENEGAL: EFFECT OF NEEM OIL AND SURROUND WP ON THE BEHAVIOR AND MORTALITY OF FEMALES. A. Balayara*, Virginia Tech, Blacksburg, VA (98)

ABSTRACT

Mango is mostly grown in Niayes (north) and Casamance (south) areas of Senegal; mango production is for family consumption, local market and exportation. Senegal ranks 5th in mango export to EU in West Africa. The mango supply chain provides 40,000 jobs. However, an economic and quarantine fruit fly, the oriental fruit fly (*Bactrocera dorsalis* (Diptera: Tephritidae)), invaded Senegal around 2004. *B. dorsalis* is causing loss of mango production of up to 60 percent. Pest management of *B. dorsalis* mostly focuses on male trapping. The objective of this work is to determine effect of neem oil and Surround WP on the behavior and mortality of females. Our hypothesis is neem oil and Surround WP may prevent oviposition of females on mango. Behaviors of females were recorded with a digital camera (Panasonic, DMC-TS6) for a period of 30 min. After 48 hours, mangoes were incubated individually in a container filled with sand for pupation of larvae. After a further two weeks, the number of pupae was recorded. In cages, female mortality was recorded. Females were attracted by the control and neem oil, while mango treated with Surround WP was less attractive. Mango used in a control provided a higher number of pupae than mango treated with neem oil and Surround; 19, 3 and 0.4 pupae respectively.

A NOVEL ATTRACT-&-KILL TECHNOLOGY FOR ORIENTAL BEETLE (*ANOMALA ORIENTALIS)* CONTROL IN BLUEBERRIES. R.J. Holdcraft* and C. Rodriguez-Saona, Rutgers University, New Brunswick, NJ (101)

ABSTRACT

A multi-year study was conducted in commercial blueberry farms to evaluate a novel attract-&-kill (A&K) technology to improve the efficacy of mating disruption (MD) for oriental beetle, *Anomala orientalis*, while lowering the amount of pheromone. Previous research showed that SPLAT™ (Specialized Pheromone & Lure Application Technology; ISCA Technologies) containing the oriental beetle pheromone could be as effective as other MD technologies. I tested the hypothesis that the addition of a toxin to the MD formulation increases its efficacy in controlling oriental beetle. In 2011, to investigate SPLAT-A&K technology for oriental beetle control, I compared the efficacy of SPLAT-MD containing 1 percent (Z)-7-tetradecen-2-one, the major oriental beetle sex pheromone component, and SPLAT-A&K containing the pheromone with additional 2 percent cypermethrin. Pheromone-baited trap captures in all plots treated with SPLAT (both MD and A&K) at 250 and 500 dollops/ha were lower compared with captures in untreated control plots, and had >95 percent disruption indices (DI). In 2014-2015, to identify lower effective pheromone amounts, I tested SPLAT-MD and SPLAT-A&K at 250 dollops/ha, but with two reduced dollop sizes: 0.25 and 0.5 g. Pheromone-baited trap captures in all plots treated with both SPLAT formulations were lower compared with captures in untreated plots; however, SPLAT-MD at 0.25 g-dollops had the lowest

DI (<90 percent). In contrast, the SPLAT-A&K treatment maintained a DI above 95 percent. These results indicate that addition of a toxin to the sex pheromone in an attract-&-kill strategy was effective at reducing oriental beetle adults in blueberries even at quantities as low as 625 grams/ha of pheromone.

HOST PLANT RESISTANCE AND PEST CONTROL: A TEST OF THE MOVEMENT-RISK HYPOTHESIS.

M.R. Scott*, Ursinus College, Collegeville, PA (102)

ABSTRACT

The potato leafhopper, *Empoasca fabae*, is a major pest of alfalfa in Northeast USA. To control this pest, farmers use pesticides that can be damaging to the environment. Two alternative pest management strategies are intercropping, in which multiple crop species are planted together rather than in monoculture, and host plant resistance, in which the crop is bred for characteristics that negatively affect the pest. Both of these strategies increase potato leafhopper movement behavior. We hypothesized that increased movement would lead to more pest-predator encounters and reduce potato leafhopper abundance. This 'movement-risk hypothesis' was tested in a mesocosm experiment with three plant treatments (leafhopper-susceptible alfalfa, leafhopper-resistant alfalfa, and leafhopper-resistant alfalfa intercropped with orchardgrass) and two predator treatments (predator present and predator absent). *Nabis americoferus* was used as the predator. As predicted by the movement-risk hypothesis, *Nabis* was more effective at reducing leafhopper abundance in the plant treatments with leafhopper-resistant alfalfa than in the plant treatment with leafhopper-susceptible alfalfa. However, intercropping leafhopper-resistant alfalfa with orchardgrass did not elevate leafhopper predation further. Our results suggest that host plant resistance may strengthen the biological control of potato leafhoppers in alfalfa.

DEVELOPING AN EFFECTIVE TRAP-AND-KILL DEVICE FOR HARLEQUIN BUG: IMPORTANCE OF TRAP COLOR AND SEMIOCHEMICAL ATTRACTANTS.

A.S. Dimeglio*, T.P. Kuhar, and D.C. Weber, Virginia Tech, Blacksburg, VA (103)

ABSTRACT

The harlequin bug (*Murgantia histrionica* Hahn) is an economically important pest of cole crops. Populations can quickly become pestiferous if unchecked. Our goal is to develop an effective trap-and-kill device for monitoring and managing populations. Current knowledge is limited on harlequin bug behavioral response to trap shape and color, and olfactory cues. To design an effective trap, we conducted a series of lab and field experiments in Blacksburg, VA to describe the behavioral response of the bug to trap architecture and color, with and without attractive volatiles.

Mixed murgantiol (harlequin bug aggregation pheromone) isomers, when combined with a regulated release of benzyl isothiocynate, were highly attractive to all life stages when combined with a tall dark-green pyramidal trap-like device. With this new knowledge we can develop a trap for harlequin bug that limits impact on natural enemies, while aiding in effective management of pest populations.

COMPARING PARASITOID PERFORMANCE BETWEEN TWO HOST PLANTS OF EMERALD ASH BORER, GREEN ASH AND WHITE FRINGETREE. J.N. Hoban*, J.J. Duan, and P. Shrewsbury, University of Maryland, Columbia, MD (104)

ABSTRACT

Since its discovery in North America in 2002, emerald ash borer, *Agrilus planipennis* Fairmare (Coleoptera: Buprestidae), an important invasive pest native to northeast Asia, has been studied aggressively in effort to control its population and dispersal. Until recently, emerald ash borer has only observed completing its life cycle, in both Asia and North America, within the ash tree genus, *Fraxinus spp*. (Oleaceae). In the summer of 2014 in Yellow Springs, Ohio, the emerald ash borer's diagnostic "D-shaped" exit holes were discovered and confirmed on white fringetree, *Chionanthus virginicus* (Oleaceae), a shrubby tree native to the southeastern United States and closely related to ash. While white fringetree is native to the southeast region of the U.S., it is commonly planted in the landscape throughout the country. In coordination with ongoing classical biological control efforts, this study aims to evaluate the ability of introduced parasitoids to detect and attack emerald ash borer in this novel host plant. Specifically, we will elucidate the effect emerald ash borer host tree species has on biological control of the established introduced parasitoid, *Tetrastichus planipennisi* Yang (Hymenoptera: Eulophidae). This study measures parasitoid oviposition on emerald ash borer larvae in choice and no-choice trials between native white fringetree, *Chionantus virginicus*, and native green ash, *Fraxinus pennsylvanica*, one of the most susceptible North American hosts. We observe the potential plant-host preference of *T. planipennisi* between the two tree species, in addition to parasitism rates and parasitoid fitness from each host. The results of this research will have implications for continued biological control of this invasive pest, as well as inform our understanding of these plant-herbivore-natural enemy interactions.

INVASIVE KUDZU BUG (*MEGACOPTA CRIBRARIA*) IN MARYLAND: THE SPREAD AND STATUS OF A RECENT INVADER. J.I. Grant* and W. Lamp, University of Maryland, College Park, MD (105)

ABSTRACT

Kudzu bugs, *Megacopta cribraria*, a known pest of soybeans, were first introduced from Japan to Georgia in 2009. Over the next four years, kudzu bugs rapidly expanded their range to their current northern limit in Maryland and have since slowed in spread. We wanted to measure the persistence of kudzu bugs from year to year in Maryland and determine the potential of cold tolerance limits on distribution. Persistence was determined by field sampling of known kudzu bugs sites from 2013 to 2015. The sampling revealed sites with kudzu bugs have continued to persist from 2013 to 2015; however, population size at these sites remains relatively low. To determine the cold tolerance of the bugs, the temperature at which their bodies freeze (super cooling point) was measured at three sites in different climatic zones during the fall of 2015. The super cooling point temperatures of the bugs ranged from -6.8 to -19.9 °C with 100 percent mortality. This range of temperatures represents an important cold tolerance parameter that can be combined with other parameters into a model for determining potential range expansion. Continued work on this project includes determining the lethal temperature for the bugs, overwintering location, and timing of movement.

PARASITOID SURVEY AND LAB DIET TRIALS OF *HALYOMORPHA HALYS* (STÃL). Z.R. Donais*, University of Connecticut, Storrs, CT (3)

ABSTRACT

Halyomorpha halys (Stål), the brown marmorated stink bug (Hemiptera, pentatomidae) poses a severe threat to agriculture in Connecticut and in the country as a whole. There is a need for detailed knowledge about successful lab-rearing techniques for the BMSB and on the native parasitoid species found attacking it in Connecticut. Thus, in order to determine the best rearing techniques, individual insects were taken from egg masses and put on seven different regularly available diet treatments and monitored daily. The diets with the best overall survival were carrot, bean carrot, and apple carrot. Apple carrot, carrot, apple, and bean carrot had the longest average survival time with detectable differences. On average, bean carrot, bean and autumn olive had the shortest periods between molts with detectable diets.

Native parasitoid populations were surveyed by using egg masses from three different stink bug species collected from lab-reared colonies, frozen at -17°C to render them unviable and placed into three different types of habitats, where they were later assessed for evidence of predator or parasitoid activity. Habitat types were a corn field, an ornamental landscape, and natural areas. Initial tests showed that natural areas had the most parasitoids detected, while ornamental plant areas showed the greatest amount of predator damage to egg masses. Two parasitoids were detected as parasitizing the stink bug egg masses, *Telenomus podisi* and *Ooencyrtus sp.* *Telenomus podisi* was detected on all three different stink bug egg masses, where *Ooencyrtus sp.* was only detected on *Chinavia halaris (Say)* green stink bug (GSB) egg masses in year one of the survey. Over 80 percent of BMSB egg masses put into the field in the ornamental plantings were predated upon, while under 40 percent of egg masses in the natural habitat were predated upon, and 0 percent of egg masses in corn were predated upon. While the greatest percent of egg masses parasitized was found in the GSB in the natural landscape where over 40 percent of the egg masses were parasitized in year one of the survey. The BMSB only had egg masses parasitized in the ornamental planting landscape during the survey.

EFFECTS OF RED CLOVER AS LIVING MULCH ON INSECT COMMUNITIES AND BELL PEPPER YIELD. H. Kahl* and C. Hooks, University of Maryland, College Park, MD (108)

ABSTRACT

Red clover is commonly used as a cover crop, but the ability of red clover to bolster wild bee populations and promote natural enemies which in turn deter pests, has not yet been extensively explored. To determine the influence of red clover on arthropod communities for bell pepper production, the abundance of herbivores, natural enemies, and pollinators in plots with bell pepper interplanted with living mulch red clover (RC) and bell pepper monoculture, or bare ground plots (BG), were compared. During the growing season, visual counts were used to assess herbivore and natural enemy populations. Slow walking transects with sweep net voucher samples were used to monitor pollinator flower visitation to red clover and bell pepper flowers. To evaluate the effect of the feeding guild structures on the pepper fruits, all fruits were assessed at the end of the season for insect damage. For RC there were more natural enemies and less herbivores early in the season; while for BG there were more natural enemies and less herbivores later in the season. The insects that caused the greatest bell pepper damage were: stink bugs and European corn borers (*Ostrinia nubilalis*). There was no significant difference between BG and RC for proportion of fruits damaged by stink bugs. However, a lower proportion of fruits were damaged by *O. nubilalis* in RC compared to BG. In addition, overall a lower proportion of fruits were damaged in RC. Pollinators rarely visited the bell pepper flowers; the primary visitors were sweat bees (Halictidae). In contrast, pollinators frequently visited the red clover flowers, especially bumble bees (*Bombus spp.) and silver-spotted skippers (Epargyreus clarus).* In summary, it seems likely that interplanted living mulch red clover could provide some pest suppression and pollinator attraction benefits for bell pepper production.

WHAT ROLE DO NATIVE CULTIVARS ('NATIVARS') HAVE IN AN ECOLOGICAL LANDSCAPE?

K.A. Nevison*, Longwood Graduate Program in Public Horticulture, Newark, DE (109)

ABSTRACT

Recent declines in pollinator populations across North America have inspired small-scale conservation efforts aimed at supporting habitat in home gardens and across communities. To supply this growing market, local nurseries often promote native cultivars ("nativars") alongside native straight species plant types, giving consumers the impression that they are equally effective in supporting pollinator habitat. Since little research has been conducted to determine the value of nativars to pollinators, I established a trial in 2015 at the Mt. Cuba Center in northern Delaware to compare *Phlox* species with its associated cultivars. Experimentation included measuring nectar volume and sucrose concentration to assess what effects, if any, these have on attracting insect pollinators. My data suggest that certain *Phlox* cultivars are superior at attracting and supporting groups of pollinators, while a majority of nativars have diminished ecological value. From these results, it is clear that more research is needed to establish specific lists of native plant cultivars which provide enhanced or similar pollinator benefits as their straight species analogues. Ultimately, this information can be supplied to nurseries and plant breeders to help promote a market for nativars that is both profitable and ecologically responsible.

WINTER MOTH *(OPEROPHTERA BRUMATA* L.*)* PARASITOID RICHNESS AND ABUNDANCE IN INFESTED AREAS IN MIDCOAST MAINE. H.D. Morin*, E. Groden, F. Drummond, C. Donahue, M. Para'k, and R.R. Kula, Department of Entomology, Smithsonian Institution, National Museum of Natural History, Washington D.C., DC (110)

ABSTRACT

Winter moth, *Operophtera brumata* L. (Lepidoptera: Geometridae), has become a serious defoliator in the environs of Harpswell, Maine. Winter moth is thought to have spread to the Maine coast around 2005 from Nova Scotia where it initially invaded in the 1930's. In the summer of 2013, *Cyzenis albicans* (Fallén) (Diptera: Tachinidae) was released in infested areas of Maine to control winter moth. Multiple factors play a role in winter moth control; the aim of this research is to characterize the diversity of parasitic wasps in areas of winter moth infestation. Data were collected from eight sites defined by the U.S. Forest Service as having "high" or "moderate" winter moth infestation based on level of defoliation risk and abundance of winter moth. Four sites were classified as high, and four were classified as moderate. Insects were collected with different sampling methods, including beat sampling of terminal tree branches, yellow pan traps, and Lindgren funnel traps. Approximately 18,378 insect specimens were collected and sorted to order; ~4,647 wasp specimens were sorted to family, and then approximately 500 wasp specimen were sorted to subfamily or genus. Specimens in taxa with species that potentially attack winter moth are currently being sorted into morphospecies. Diversity data for wasps potentially parasitic on winter moth will be analyzed using RStudio to discern levels of diversity relative to levels of winter moth infestation.

BEE DIVERSITY AND POLLEN FORAGING SPECIFICITY IN CULTIVATED HIGHBUSH BLUEBERRY (ERICACEAE: *VACCINIUM CORYMBOSUM* L.) PLANTINGS IN RHODE ISLAND. Z.D. Scott* and S.R. Alm, University of Rhode Island, Kingston, RI (111)

ABSTRACT

Bees are some of the most economically and ecologically important pollinators, necessary for the production of about one third of all crops. One such crop is highbush blueberry, grown throughout Rhode Island in small acreages. We conducted a survey of the native bee species foraging on managed blueberry farms throughout Rhode Island, and analyzed their preference for blueberry pollen. We mostly found species in the genera Bombus and Andrena. We found that A. bradleyi collected the highest percentage of blueberry pollen (97.2 percent ±1.7 SE). Total grains collected was related to the size of the bee, with the large species B. grisecollis collecting the most (5.74e5 ±9.74e4). This is likely due to the fact that Bombus queens are highly active during blueberry bloom. We conclude that for growers to obtain the greatest benefit from bees, they should focus forage and nesting habitat management efforts on the few species found most frequently.

OVER THE SOIL, AND THROUGH THE GUT, TO CUCURBIT HEALTH WE GO!: POTENTIAL BIOLOGICAL CONTROL OF *ERWINIA TRACHEIPHILA* BY *PSEUDOMONAS FLUORESCENS* VIA CUCUMBER BEETLE GUT INTERACTIONS.
D.C. Roberts*, S. Fleischer, J. Sakamoto, and J. Rasgon, Penn State University, University Park, PA (113)

ABSTRACT

Erwinia tracheiphila is the causal agent of bacterial wilt of cucurbits and is transmitted by the striped cucumber beetle, *Acalymma vittatum.* Management of this disease and beetle increase insecticide inputs in the Northeastern United States. The causal agent of fire blight of tree fruit, *Erwinia amylovora,* has a commercially available biocontrol from common soil bacterium, *Pseudomonas fluorescens*, which uses competitive exclusion to decrease disease occurrence. Based on biological similarities of *E. tracheiphila* and *E. amylovora,* we hypothesized that *P. fluorescens* should competitively exclude *E. tracheiphila.* To test for potential inhibition, *E. tracheiphila* and *P. fluorescens* were cultured on King's B medium using a disk diffusion assay. The plates were monitored daily for growth of the bacteria and zone of inhibition production. Our *in vitro* experiments illustrated that *P. fluorescens* does inhibit *E. tracheiphila*. The next step was to visualize *P. fluorescens* and *E. tracheiphila in vivo* in the beetle gut to determine co-localization and interaction potential. This was done by fluorescent *in situ* hybridization of beetle guts after bacterial inoculation utilizing five treatments. The *in vivo* experiments provided strong evidence of *P. fluorescens* limiting *E. tracheiphila* presence via beetle gut interactions.

RESEARCH POSTERS

COMPARING THE USE OF AGGREGATION PHEROMONE LURES WITH TRADITIONAL VISUAL SAMPLING OF *HALYOMORPHA HALYS* IN NEW JERSEY PEACH ORCHARDS.
W.P. Woodroffe*, A.L. Nielsen, and B.R. Blaauw, Rutgers University, New Brunswick, NJ (114)

ABSTRACT

Halyomorpha hayls, commonly known as the Brown Marmorated Stink Bug, is an invasive agricultural pest of North America causing particularly heavy losses in Northeastern tree fruits. Currently its populations are monitored using visual sampling methods and more frequently now using traps containing its recently discovered aggregation pheromone.

In this experiment, two commercially available aggregation pheromone lures were tested in conjunction with visual tree samples and compared to absolute counts, done by observing post-spray kills, to see which method was a more reliable indicator of population size. The experiment took place at Rutgers Agricultural Research and Extension Center in Bridgeton, New Jersey during the 2015 growing season.

A NEW LURE FOR *FRANKLINIELLA TRITICI* (THYSANOPTERA: THRIPIDAE). D. Teulon, M. Nielsen, S. Jones, and T.C. Leskey*, USDA-ARS, Kearneysville, WV (115)

ABSTRACT

Frankliniella tritici (Fitch) or eastern flower thrips (EFT), is widespread in North America, is polyphagous and associated with a wide range of species of flowering plants. EFT causes damage to various field crops but is not known to transmit plant viruses. Two trials were undertaken to determine the effectiveness of the non-pheromone thrips lure, methyl isonicotinate (MI), at the Appalachian Fruit Research Station, WV, USA in June and July 2015, respectively. Trapping bioassays involved white 2-L volume water traps in grass fields with and without MI dispersers. In trial #1 (10 m x 10 m matrix of 4 treatments (3 MI dispensers and control) and 4 replicates, 48 hr duration) MI increased trap catchers of EFT by about 2x. In trial #2 (paired traps (one MI dispenser and control) 20 m apart x with replicates 400 m apart, 24 hr duration) MI increased trap capture of EFT by about 40x. Daily average temperatures were around 30–35°C in both trials. This research confirms the efficacy of MI, the active ingredient of LUREM-TR, as an attractant for EFT, and extends the number of thrips species known to respond to this semiochemical to 11. This research further underpins the view that baited traps influence trap capture of neighbouring traps and that previous work might have underestimated the effect of semiochemicals to increase trap capture, especially where there was a high concentration of baited traps in a given area.

HOW DESTRUCTIVE IS BROWN MARMORATED STINK BUG TO HERBACEOUS PERENNIAL PLANTS. S. Gill*, B. Kunkel, D. Smith-Fiola, K. Rane, V. Rosenkranz, and S. Klick, University of Maryland, College Park, MD (118)

ABSTRACT

The brown marmorated stink bug (BMSB) is rapidly expanding its territory and this invasive species has developed a palate for many species of plants. The pest has spread to more than 40 states where it has been found feeding on a number of species not recorded in its native habitat in Asia. Furthermore, their populations are frequently numerous enough to cause significant economic injury to many agricultural crops, but little is known about economic threats to ornamental plants. BMSB adults have been observed feeding on flowers and seed pods of several ornamental herbaceous plants, but the damage to flowers, seed pods, stems and leaves has not been documented. Disease transmission from BMSB feedings on herbaceous plants has not been confirmed in the literature. Because of this dearth of knowledge, our team of researchers launched a study to determine if BMSB was a significant pest of herbaceous perennial plants growing in commercial perennial plant production nurseries. Trials were conducted at nurseries in Maryland and Pennsylvania in 2012. Our results show limited damage to herbaceous perennials. A list of potential target herbaceous perennials has been developed.

DIGITAL BOOK FOR WEED IDENTIFICATION. B.A. Ackley*, The Ohio State University, Columbus, OH (120)

ABSTRACT

Plant identification can be challenging and even intimidating for the inexperienced. Growers do not necessarily need to identify every weed in a field to be effective managers, but should be able to identify the major weeds that are important to their operations and goals. At first glance, learning how to identify weeds can seem like a daunting task given the number and diversity of species, but it is not as difficult as it may seem. Generally, there is a specific group of weeds that tends to dominate disturbed habitats within any native landscape. This iBook, "The Ohio State University Guide to Weed Identification", was created to help people better understand the nature of the weeds they are trying to control, and plant identification is a key component of that understanding. The iBook provides a new way to use an old tool - visualization - in the world of weed identification. Plant descriptions contained herein include key identification characteristics, photos of many species at different stages of maturity, and 360-degree movies for most species in the book. This book is not meant to be a compendium of all weedy plants in the U.S., but rather includes a number of the most common Midwestern U.S. weeds and the basic intellectual tools that are necessary to successfully identify plants.

HISTORICAL PERSPECTIVE OF THE NORTHEASTERN BRANCH ASA-CSSA-SSSA. M. Fidanza*, Pennsylvania State University, Reading, PA (121)

ABSTRACT

The Northeastern Regional Branch of the American Society of Agronomy, Crop Science Society of American, and Soil Science Society of America (or ASA-CSSA-SSSA or "tri-societies") provides a forum for the interaction of persons interested in all aspects of agronomic, crop, and soil sciences and management. The Northeastern Branch ASA-CSSA-SSSA comprises the states of Connecticut, Delaware, Massachusetts, Maryland, Maine, New Hampshire, New Jersey, New York, Pennsylvania, Rhode Island, Vermont, Washington DC, West Virginia, and the Canadian Provinces of Newfoundland/Labrador, Nova Scotia, Ontario, Prince Edward Island and Quebec. The ASA was formed in 1907, with the primary objective to "Increase the dissemination of knowledge concerning soils and crops and the conditions affecting them". The SSSA was formed in 1936, and the CSSA was formed in 1955. Soon after that, all three societies merged to form the "tri-societies" of the ASA-CSSA-SSSA. The regional branch was originally the Northeastern Branch ASA only, and is now officially named the Northeastern Branch of the ASA-CSSA-SSSA. Northeastern Branch meetings are typically held once per year and at times every other year in June or July, and the locations have been at universities and colleges within the geographical region. The joint 2016 Northeastern Plant, Pest, and Soils Conference will provide a venue for sharing ideas and the education of members in the fields of agronomy, plant pathology, entomology, horticulture, and weed science, and to provide outreach to practitioners and other professionals associated with those aforementioned disciplines.

SEASONAL VARIATION OF LITTORAL ZONE BENTHIC MACROINVERTEBRATES IN COMPARATIVE STUDY OF FOUR LAKES IN SOUTHEASTERN NEW YORK. T.M. Cloherty* and J.W. Rachlin, Penn State University Abington, Abington, PA (123)

ABSTRACT

In this study, we examined four lakes in suburban Westchester County, New York from April through October 2009 and April through October 2010 to determine if species assemblages of benthic macroinvertebrates in the littoral zone demonstrated seasonal variation. Species richness, relative abundance and diversity were examined by season for each lake and between the lakes. In addition, physicochemical factors were also examined to determine if there was any significant variation among means of physicochemical factors in the lakes throughout the seasons. One of the lakes was in separate hydrological drainage, which we hypothesized would result in differences in diversity and species richness. Contrary to our hypothesis, we found no significant differences in species diversity or in physicochemical variables between the lakes. We did detect seasonal patterns in species assemblages, which we determined through multivariate models.

DEVELOPMENT OF A REGIONAL POLLEN LIBRARY TO INVESTIGATE FORAGING PATTERNS IN *APIS MELLIFERA*. J.I. Wolfin*, University of Delaware, Newark, DE (125)

ABSTRACT

One of the critical components to reestablishing healthy, sustainable bee colonies in the United States is ensuring that bees have suitable sources of nutrition, specifically pollen. Pollen can vary in nutritional quality depending on the percent crude protein, so it is critical that bees obtain forage from floral sources that have pollen grains with high amounts of protein. In order to determine the nutritional quality of a pollen grain, it is necessary to be able to identify a pollen grain. Pollen grains were collected from four sites in the Mid-Atlantic to create a fully functional pollen library, in partnership with Discover Life. Grains were analyzed according to a variety of morphological characteristics to ensure that users would be able to identify grains to the lowest taxonomic rank possible. Bee-collected pollen from The University of Delaware Botanical Garden (UDBG) in Newark, Delaware and The Mount Cuba Center (MCC) in Hockessin , Delaware were examined to determine which types of pollen were most prevalent. While more pollen was collected at MCC, it was found that differences in foraging totals were not statistically significant. Analysis of bee-collected pollen suggested that bee foraging patterns do change temporally, as different species/genera became prevalent at different time periods. This research warrants further investigation to determine the nutritional quality of the most prevalent pollen grains, and shows the need for an extensive pollen library.

PREVENTION OF BRASSICA CROP LOSSES FROM NEW INVASIVE SPECIES, SWEDE MIDGE (*CONTARINIA NASTURTII* KEIFER) ON AT-RISK SMALL-SCALE ORGANIC FARMS. C.A. Hoepting* and C.A. Hall, Cornell University Cooperative Extension, Albion, NY (126)

ABSTRACT

Swede midge (*Contarinia nasturti* Keifer) (Diptera; Cecidomyiidae) (SM) is an invasive pest of brassica crops in North America with a rapidly expanding range. SM larval feeding causes swelling and distortion of meristematic plant tissues, resulting in up to 100 percent yield loss of broccoli, Brussels sprouts and kohlrabi. Since SM was first identified in Ontario, Canada in 2000, it has spread throughout southeastern Canada and to northeastern regions of the US including New York. Currently, there are no effective organic methods for SM. The objectives of this study were to 1) describe the activity of SM in brassica fields as it related to management practices, and 2) evaluate efficacy of a combination of insect-exclusion netting and various mulch types.

To address the first objective, SM populations were monitored using female sex pheromone traps on six small-scale organic farms located in five counties in New York in 2015. Traps were deployed at 59 sites where brassica crops were grown and monitored weekly for male SM. Crops were assessed for SM damage at harvest. To address the second objective, insect exclusion netting and various mulches including black plastic, landscape fabric, hay and straw were evaluated for protecting brassica crops from SM damage. This experiment was conducted on four farms that had natural SM infestations. Treatment plots were 4-ft wide with a minimum of 20 plants per replicate and treatments were replicated 1-2 times. Traps were placed in each treatment to monitor SM population and crop damage was evaluated at harvest.

Results from the first experiment revealed that emergence of the overwintering SM population began during the first half of May. It was bimodal with the first peak occurring in mid-June and the second of similar magnitude occurring during the last week of June. Dates of first emergence and peak flights, and magnitude of peak flights varied slightly and considerably, respectively, from farm to farm and from field to field on the same farm. In secluded fields, more SM damage occurred in red cabbage that was planted during spring emergence in May (damage rating; 2.8/4.0 = severe) than when it was planted after spring emergence was complete in August (damage rating: 0.6/4.0 = minor). Results from the second experiment demonstrated that insect-exclusion netting was highly effective in protecting the crop from SM damage. No broccoli was damaged when using the netting, whereas 94-100 percent of the broccoli was damaged and 50-85 percent unmarketable in the untreated control (=no netting). Use of mulch under the netting reduced SM damage by 66 percent compared to bare ground when it was used on top of a SM emergence site. Most importantly, weeds were managed more effectively using all mulch types compared with bare ground. Differences in plant development and quality occurred between netting and open air and among mulch types under the netting that resulted in both increased and decreased yield. Also, netting provided beneficial exclusion of flea beetles and detrimental inclusion of cabbage worms and slugs.

Overall, crop rotation/separation can be a very effective management strategy for SM, but timing plantings to avoid SM emergence is critical. The use of insect-exclusion netting and mulches will protect brassica crops from SM, but its use regarding its effects on plant development and quality, and entire pest complex needs to be optimized.

SEEDBANK DYNAMICS OF INVASIVE SWALLOWWORTS (*VINCETOXICUM* SPP.).
A. DiTommaso*, L. Milbrath, S.H. Morris, C.L. Mohler, and J. Biazzo, Cornell University, Ithaca, NY (127)

ABSTRACT

Pale swallowwort (SW) (*Vincetoxicum rossicum*) and black swallowwort (*V. nigrum*; Apocynaceae, subfamily Asclepiadoideae) are European viny milkweeds that have become invasive in many habitats in the northeastern U.S.A. and southeastern Canada. A multi-year seed bank study was initiated in fall 2011 to assess annual emergence and longevity of seed of pale SW and black SW at four different burial depths (0, 1, 5, and 10 cm) over four years. The experiment was conducted in pots buried with the rim nearly flush with the surrounding soil and filled with a locally collected soil. One hundred swallowwort seeds were sown in seed pans buried in each pot. Initial seed viability was 97 percent (black SW) and 95 percent (pale SW). Pots were checked twice weekly for swallowwort seedling emergence beginning in early May and weekly from July through September. Emerged seedlings were removed. Beginning in October 2012, seed pans were retrieved annually and recovered seed, seed coats and seed coat fragments were counted. Filled seeds (i.e., that appeared to have an embryo) were tested for viability with a 1.0 percent solution of tetrazolium chloride. The majority of seedling emergence occurred during the first year (92 percent in 2012), and no new seedlings emerged in the third (2014) or fourth (2015) years. During the 2012 growing season, pale SW had relatively poor emergence at the 0 cm (12 percent), 5 cm (7 percent), and 10 cm (0.05 percent-only one seedling) sowing depths, while at 1 cm 38 percent of pale SW seeds emerged. The larger seeded black SW was more successful, with two-thirds of all sown seeds emerging at the 1 cm (71 percent) and 5 cm (66 percent) depths and 27 percent emerging at 10 cm. Only 17 percent of the surface-sown black SW emerged. A large portion of the seeds that germinated in the first year at 5- or 10-cm depths died before reaching the soil surface (pale SW-97 percent, black SW-48 percent). Of filled seeds that were recovered in 2012, mainly black SW at the 0 cm depth, 66 percent were viable. No viable seeds were recovered after the second growing season. Seeds recovered following the third year had become too deteriorated to accurately assess. Key findings are that swallowwort seeds do not appear to survive more than two years in the soil, at least in our experiment, and that the two swallowwort species can germinate and emerge from soil depths of 10 cm which was not expected.

RANGE EXPANSION OF THE INVASIVE PEST *MACONELLICOCCUS HIRSUTUS* IN THE MAJOR COCOA PRODUCING REGION OF SOUTH AMERICA.
M.P. Culik*, D. dos S. Martins, C. A. S. Souza, M. J. Fornazier, and J. S. Zanuncio, Instituto Capixaba de Pesquisa, Assistancia Tecnica e Extensao Rural (INCAPER), Vitoria, Brazil (128)

ABSTRACT

Maconellicoccus hirsutus (Green) (Hemiptera: Pseudococcidae) is an invasive mealybug that has become established in almost all biogeographic regions. This insect is a potential pest of many crops and native plant species. *Maconellicoccus hirsutus* was first detected in Brazil in 2010 in the northern region of the country. This study was conducted to determine the distribution of *M. hirsutus* and document the recent range expansion of the pest on cocoa (*Theobroma cacao* L., Malvaceae) in major production regions of Brazil. A total of 460 cocoa plantations were sampled over a period of one year (2013 - 2014). *Maconellicoccus hirsutus* was found on cocoa of farms in 8 municipalities of Bahia, and in 13 municipalities of Espírito Santo. This is the first report of *M. hirsutus* on cocoa in Brazil and this study documents the recent invasion of the major cocoa producing region in South America by *M. hirsutus.* Integrated pest management is required to address the increasing problem of invasive pests such as *M. hirsutus*. Introduction of host-specific parasitoids of *M. hirsutus* where they are currently not present (classical biological control) should be considered to reduce negative impacts of this pest on economically and environmentally important plant species.

EFFECT OF WATER AVAILABILITY ON COMPETITIVE INTERACTIONS BETWEEN TWO INVASIVE SWALLOWWORTS (*VINCETOXICUM* SPP.) AND COMMON MILKWEED. M.C. Joline and A. DiTommaso*, Cornell University, Ithaca, NY (129)

ABSTRACT

The impacts of climate change on water availability in natural systems are becoming apparent. The increasing frequency and intensity of drought conditions can potentially alter plant growth and thus competitive interactions between nonnative invasive plant species and native species. Pale swallowwort (SW) (*Vincetoxicum rossicum*) and black swallowwort (*V. nigrum*; Apocynaceae, subfamily Asclepiadoideae) are European perennial herbaceous vines that have become invasive in many habitats in the northeastern U.S. and southeastern Canada. Common milkweed (*Asclepias syriaca*) is a native species commonly found in the same habitats invaded by these two invasive vines. A greenhouse study was conducted for 12 weeks to determine the competitive interactions between each of the swallowwort species and common milkweed under adequate and limiting water conditions. Pale SW and black SW were each grown at five replacement series densities with common milkweed in a randomized complete block design with 7 replicates. Plants were grown in 11.4 L pots with a total density of four plants per pot in proportions of 4:0, 3:1, 2:2, 1:3, and 0:4, pale SW or black SW: common milkweed. Each set of five replacement-series density pots was subjected to either adequate watering or limited watering. At the start of the trial, all pots were watered daily for 20 days to ensure seedling establishment. After this period, the adequate watering treatment pots were watered daily to field capacity, while the limited watering treatment pots received no water during five separate 4-day drought-periods in the 12-week experiment. The soil moisture content of all pots was recorded daily during each of the five drought periods. At harvest, the average height and aboveground dry biomass of plants in each pot were recorded. The watering treatment and plant density ratio differed in their effect on average height of the two invasive vines. Higher density of pale SW resulted in an increase in average height, but a similar result was not observed for black SW or common milkweed. Pale SW height under adequate watering (29.7 ± 4.8 cm) was greater relative to the limiting water treatment (21.8 ± 3.2 cm). However, height of black SW did not differ between the adequate (47.3 ± 4.6 cm) and limiting (40.8 ± 6.6 cm) watering treatments. The height of common milkweed plants under limiting water (17.3 ± 1.1 cm) was also reduced relative to that under adequate water (19.9 ± 1.3 cm). Pale SW aboveground biomass did not differ between the adequate (2.45 ± 0.40 g) and limiting (1.73 ± 0.25 g) water treatments. However, black SW biomass was greater in the adequate water (3.69 ± 0.41 g) versus limiting water (2.57 ± 0.27 g) treatment. Nonetheless, the biomass of black SW was greater than the biomass of pale SW under both the adequate and limiting watering conditions. These preliminary findings suggest that black SW may be more adapted to periods of drought or other limiting water conditions than PSW or common milkweed and represents a greater competitive threat to common milkweed than pale SW.

ENHANCING POLLINATOR HABITAT IN NEW ENGLAND. A. Papineau*, C. Neal, and R. Maccini, University of New Hampshire, Durham, NH (4)

ABSTRACT

Pollinators rely on access to abundant supplies of flowering plants from which to forage pollen and nectar for sustenance. Abundance and diversity of native bees as well as honeybees and other pollinators is supported by conserving and planting wildflowers and other forage on farms and in the landscape. Modifying management practices such as mowing and pesticide strategies also helps support pollinator health. Members of the Northern New England Pollinator Habitat Working Group (NNEPH-WG) support enhancement of pollinator habitat in the region through research, education, and demonstration. This poster provides an overview of selected pollinator habitat enhancement projects led by NNEPH-WG members.

POST-PLANTING BIOLOGICAL MANAGEMENT OF APPLE REPLANT DISEASE. T.L. Bradshaw*, University of Vermont, Burlington, VT (134)

ABSTRACT

Management of apple replant disease (ARD) after orchard planting with biopesticides is a relatively new practice. However, causative agents of ARD vary by orchard location, so specific biopesticide materials may vary in efficacy for any given orchard. Research was conducted in 2014 and 2015 to evaluate effects of commercial biopesticides on tree growth, crop yield, and nematode populations in two ARD-affected Vermont orchards planted in 2011. The first orchard is located at the University of Vermont Horticulture Research & Education Center in South Burlington, VT (HREC) with Windsor Adams loamy sand soil. A previous orchard was maintained on the planting site from 1990 through 2009. 'Royal Empire' trees grafted to Budagovsky 9 dwarfing rootstocks were planted in 2011 in a tall spindle training system with tree density of 2990 trees*ha^{-1}. The second planting is located at a commercial orchard in South Hero, VT (SHVT) on Amenia and Kendall silt loam soils. The previous orchard was maintained on the site from 1900 to 2009. 'McIntosh' trees on semidwarf EMLA-26 rootstock were planted at a tree density of 852 trees per acre

At each site, three treatments were applied to five replicates per treatment in a completely randomized design. At HREC five-tree replicates and at SHVT two-tree replicates were used. Treatments included: 1) non-treated (water) control (NTC); 2) Actinovate AG at 0.84 kg*ha^{-1} (ACT); and 3) MeloCon at 4.4 kg*ha^{-1} (MCN). Treatments were applied as a soil injection with Rears Pak-Tank Sprayer (Rears Mfg Co., Eugene, OR) and OESCO FN-12 Root Feeding Needle (OESCO, Inc., Conway, MA) within the drip line of individual trees in the equivalent of 1800 and 1500 l*ha^{-1} at 689 kpa at HREC and SHVT, respectively. ATreatments occurred on four dates from May through July for each site in each year.

Measured paratmeters included central leader and vegetative shoot growth, trunk cross sectional area (TCSA), canopy height and width (2015 only), number and kg fruit yield, and total nematodes per g soil, identified to genus (*Meloidogyne and Pratylenchus*). Data were analyzed within orchard and year by ANOVA (SAS 9.3) with Tukey's adjkustment at α=-0.05. Few differences were observed among the treatments for measured parameters. At HERC, ACT trees had greater TCSA than NTC in both years, but no differences in TCSA were observed at SHVT. In 2014, ACT-treated trees at HERC had greater leader length than NTC. Tree canopy width was greater on ACT than NTC at HERC in 2015. No differences were observed in crop yield or nematode density in either orchard.

ALLELOCHEMICALS FROM CEREAL RYE FOR WEED MANAGEMENT. B.A. Otte*, S. Mirsky, C.P. Rice, K. Tully, and H. Schomberg, University of Maryland, College Park, MD (135)

ABSTRACT

With the increase in multiple herbicide resistant weeds and the challenge in managing weeds in organic systems, there is a need to improve multi-tactic weed management. Cover crops have played an important role in development of multi-tactic weed management programs with an emphasis on physical weed control and to a lesser extent, phyotoxcity from allelochemicals released during decomposition. While allelochemicals have not been shown to play a primary role in weed suppression in either system, to best exploit this tactic, it is necessary to better understand the interactions between allelochemicals and other weed control tactics. Furthermore, little work has been done to examine the interactions between cover crops, the soil rhizosphere, and bioavailability of these allelochemical compounds (benzoxazinoid and phenolics) over time. Therefore, we conducted a study that examines the presence of these chemicals in the soil and the decomposition of these allelochemicals over time. Preliminary results suggest that the concentrations of these compounds are prevalent in three growth stages of the rye, with the concentration varying over time. These results will enhance the understanding of multi-tactic weed management and allow for better comprehension of allelochemicals in weed management.

EVALUATION OF DIFFERENT COVER CROPS IN INTERSEEDED CORN. K.E. Caswell*, W.S. Curran, S. Mirsky, G. Roth, M.R. Ryan, and J.M. Wallace, Penn State, University Park, PA (136)

ABSTRACT

Cover crop adoption in the Northeast has been slow due to a number of factors. One reason is that farmers are faced with a limited window of for cover crop planting following grain harvest and before weather conditions prevent fall growth. Multiple methods have been evaluated to expand this restricted planting window including aerial seeding, underseeding and relay intercropping. All of these methods face additional challenges in no-till production systems. Researchers at Penn State developed an innovative cover crop interseeder to address establishment challenges in no-till grain production. After design of the machine was completed, potentially successful cover crops were evaluated in no-till grain. A successful interseeded cover crop may require traits that are not normally associated with post-harvest established cover crops. These species need to tolerate low light and potentially low moisture conditions while they germinate, establish, and grow. Annual ryegrass [*Lolium perenne*, L. ssp. *multiflorum* (Lam.) Husnot] has been shown to be a strong candidate for interseeding in earlier studies and a trial was conducted to test different varieties for interseeding success. A Species Trial was also conducted to evaluate ten different grass and legume species. The Species Trial was conducted in Pennsylvania (PA), New York and Maryland (MD). The Annual Ryegrass variety trial was conducted only in PA and MD. Several species were shown to be better candidates for interseeding based on fall and spring biomass. These species were: medium red clover (*Trifolium pratense* L.), annual ryegrass, and orchardgrass (*Dactylis glomerata* L.). The Annual Ryegrass variety trial showed similar performance across varieties with few performing better than others. A better understanding of appropriate cover crop selection can be used in conjunction with ongoing work in herbicide trials to create recommendations for farmers to improve the effectiveness of interseeding cover crops in grains.

NO-TILL ORGANIC COVER CROP PERFORMANCE AND WEED SUPPRESSION IN PENNSYLVANIA SMALL GRAINS. T.R. Mazzone*, W.S. Curran, and J.M. Wallace, Pennsylvania State University, State College, PA (137)

ABSTRACT

Weed management and cover crop establishment are significant challenges in rotational, no-till organic grain production. A rotational no-till organic grain systems experiment was initiated at the Penn State Rock Springs Agronomy Farm in 2014 using a corn-soybean-spelt rotation to assess alternative strategies for integrating cover crops and reducing tillage. One strategy is to reduce the intensity of tillage during the establishment of the spelt following soybean. In addition, tillage frequency may be reduced by under-seeding a legume cover crop into the winter grain using a no-till grain drill, which extends the growing season of the legume before corn the following year. Farmer adoption of this strategy will depend on spelt performance as well as the legume cover crop and subsequent corn yield. Within our cropping systems study, we evaluated the effects of the spelt and cover crop establishment method on spelt performance and weed abundance. Establishment methods included: 1) Full inversion tillage prior to spelt seeding and no cover crop, 2) Minimum tillage (chisel) prior to spelt seeding and no cover crop, 3) Inversion tillage prior to spelt seeding followed by under-seeded medium red clover (12.3 kg ha-1) + timothy (3.4 kg ha-1) in late March, and 4) No-till drill seeding of spelt followed by the under-seeded mix. Liquid dairy manure was applied to all systems prior to spelt seeding through broadcast or subsurface injection in the no-till system. In 2015, all four cropping system treatments had similar spelt yield (3,093 kg ha-1) that ranged from 2,240 – 3,644 kg ha-1. Under-seeded cover crop biomass after spelt harvest was lower under no-till management (759 kg ha-1) than in inversion-tilled systems (951 kg ha-1). The most common problem weeds in all treatments included foxtail species (57-80 percent of total weed biomass) and common ragweed (8-20 percent). Overall, weed biomass was higher in no-till and reduced-till systems (416 kg ha-1) compared to inversion tillage systems (161 kg ha-1). The use of inversion tillage with an under-seeded cover crop resulted in the fewest (73 kg ha-1) annual grass weeds, while no-till (256 kg ha-1) and minimum tillage (323 kg ha-1) plots without a cover crop had greater annual grass weed biomass. Common ragweed was a major weed of no-till (71 kg ha-1) treatments, compared to inversion tillage and a cover crop (14 kg ha-1). The positive effects of under-seeded legume cover crops on spelt yield and weed suppression will continue to be evaluated in 2016-17. These data can inform better strategies for adoption of no-till spelt as a productive rotational cash crop.

THE USE OF PHOTOSYNTHETIC FLUORESCENCE MEASUREMENTS AS A DIAGNOSTIC TOOL TO IDENTIFY GLYPHOSATE RESISTANT GIANT RAGWEED (*AMBROSIA TRIFIDA*) BIOTYPES. K. Segobye*, R. Robertson, S.C. Weller, and B. Schulz, University of Maryland, College Park, MD (139)

ABSTRACT

A popular option for evaluating injury due to herbicides is by visual scoring. This results in a subjective measurement often complicating accurate comparisons between different experiments from separate publications. A second option is to assess fresh weight or dry weight or a ratio of the two. This provides quantitative data for comparison between experiments. However, its convenience and bias is debatable considering response to herbicides such as glyphosate is not an immediate response resulting in long-term experiments or premature termination of an experiment in which the entire susceptible plant is not dead. A third option is to measure only the tissue growth after herbicide application. This also introduces bias due to the difficulty in determining defined re-growth.

Here we propose the use of chlorophyll fluorescence measurements for comparison between glyphosate sensitive and resistant giant ragweed after herbicide treatment as rapid and reliable mean to measure herbicide response in resistant and sensitive plants. The measurements of measuring the activity of PS II reaction centers resulting in a quantum yield of converted energy. These measurements are often used to assess plant stress prior to the development of visual symptoms. Emission of fluorescence energy from radical pairs of PSII and PSI returning to ground state after photon excitation was measured using an Imaging-PAM (Pulse Amplitude Modulation) instrument on treated and untreated, glyphosate sensitive and resistant giant ragweed plants. We show the quantitative relationship between fluorescence measurements and glyphosate resistance as a mean to rapidly identify glyphosate resistant plants in this weed.

VEGETATIVE PROPAGATION OF *AMBROSIA ARTEMISIIFOLIA* FOR RAPID RESISTANCE TESTING. B.W. Schrage* and W.J. Everman, North Carolina State University, Raleigh, NC (140)

ABSTRACT

During the summer of 2015, soybean producers in Northeastern North Carolina began experiencing insufficient control of common ragweed (*Ambrosia artemisiifoia*) with the use of labeled PPO-inhibiting herbicides. The existence of confirmed PPO-resistant common ragweed in Delaware and Ohio prompted an immediate investigation. Distance from experimental stations and the documented difficulty associated with screening common ragweed from seed collected previously in the same year created a need for a simple, albeit effective, resistance screening method.

Live samples were collected, transplanted, propagated and an accelerated resistance screening was conducted. A completely randomized design including five herbicide treatments (fomesafen, fomesafen+glyphosate, lactofen, aciflourifen at respective full labeled rates and an non-treated check), two putative resistant populations (Moyock and Sunbury, NC), and five replications was employed at the Method Greenhouse facilities in Raleigh, NC.

Significant differences in plant height and injury were observed among herbicide treatments ($P > 0.0001$) although all plants survived and showed potential to recover. Numeric trends suggest that lactofen had the greatest toxicity on specimens; whereas acifluorfen had the least. Sample locations in Moyock and Sunbury, NC both suggest the existence of PPO-resistant common ragweed; although more detailed screening methods need to be conducted. Initial results from this research raise concerns that PPO-Inhibiting herbicides may become ineffective should over-reliance continue in North Carolina soybean production.

GIVEN HIGHER YIELD POTENTIALS, IS IT TIME TO CHANGE OUR NITROGEN RECOMMENDATIONS FOR CORN? S.N. Swink*, Q.M. Ketterings, K. Czymmek, G.S. Godwin, K. O'Neil, M. Hunter, E. Bever, M. Contessa, J. Lawrence, D. Morse, T. McClelland, P.M. Barney, and A.C. Tagarakis, Cornell University, Ithaca, NY (141)

ABSTRACT

In New York and many other states, corn silage and grain yields have increased at a fairly consistent rate over the past 65 years. Knowing the yield potential (YP) for corn on individual fields is important to management of nutrient inputs as yield is a critical component of calculations used to determine recommended nitrogen (N) fertilizer and manure application rates. At Cornell University, the land grant university of New York, a database with YPs for corn was developed. In this database, YP is soil type and artificial drainage specific, and reflects an average yield of 4 out of 5 crop years under excellent management. The database was last updated in the 1990s which raises two questions: (1) With gains in corn genetics and overall crop production, should the corn yield potentials that drive N recommendations be re-evaluated? (2) Does higher productivity mean more N needs to be supplied through manure and/or fertilizer, or are new varieties simply making better use of existing N? In 2013 and 2014 as part of an on-going study using a new, on-farm adaptive management approach, we evaluated yields of 36 farmer-selected 2nd year or higher corn silage or grain fields. Data were collected for 2 to 4 fields/farm, across 19 soil types representing all 5 soil management groups in New York (from clay to sandy soils). The average yield across all 36 fields equaled 113 bu/ac (19.2 ton/acre silage), slightly less than the Cornell University listed average YP of 117 bu/ac (20.9 ton/ac silage) for the soil types in the study. However, 39 percent of the study fields yielded less than 90 percent of the listed YP, 36 percent of the fields were within 10 percent, while 25 percent of the fields yielded greater than 110 percent of the listed value. The average yield for brown mid-rib (BMR) varieties was lower than the average yield of conventional corn varieties across the two years. Nitrogen applications to fields that yielded below the Cornell University reported YP were generally equal to or greater than crop N removal (10 of 12 fields), while the highest yielding corn fields and two-thirds of those with yields within 10 percent of the Cornell YP removed up to 60 percent more N on a pound for pound basis than was applied. These results suggest that the highest yielding sites make most efficient use of N additions (needing less external N per bu or ton of silage produced) and highlight the importance of individual field assessments for monitoring yields and determining optimal N application needs.

GLYPHOSATE-RESISTANT WEED CONTROL AND SOYBEAN TOLERANCE TO PPO-HERBICIDES APPLIED AT TWO DIFFERENT GROWTH STAGES. J.S. Aulakh*, P.S. Chahal, and A.J. Jhala, Connecticut Agricultural Experiment Station, Windsor, CT (142)

ABSTRACT

With the increasing incidence of multiple herbicide-resistant weeds, weed management has become a serious challenge for soybean growers. In Nebraska, 5-enolpyruvyl shikimate-3-phosphate synthase (EPSPS) as well as acetolactate synthase (ALS) -inhibitor herbicide-resistant weeds occur in many soybean fields where herbicides from these modes-of-action have been repeatedly used in the past. Greenhouse experiments were conducted in 2014 to evaluate the efficacy of PPO-inhibitor herbicides applied POST for the control of three glyphosate-resistant (GR) weeds and potential for soybean injury, when applied at two growth stages. All herbicide treatments controlled 10- and 20-cm tall GR common waterhemp ≥ 95 percent at 21 DAT. GR giant ragweed and kochia were controlled 86 to 99 percent when treated at 10-cm height and 78 to 92 percent at 20-cm height. Herbicide treatments reduced shoot biomass in the three GR weeds 88 to 100 percent when treated at 10-cm height and 73 to 100 percent when treated at 20-cm height, at 21 DAT. Soybean injury and shoot biomass data revealed that acifluorfen and lactofen were more injurious (≥ 17 percent), whereas fomesafen, and fomesafen plus glyphosate were relatively safer (< 10 percent injury). Overall, fomesafen and fomesafen plus glyphosate caused the least injury to soybean and were more effective in controlling GR common waterhemp, giant ragweed, and kochia compared with acifluorfen and lactofen.

HERBICIDE BANDING IN CORN AND ASSOCIATED FLORAL DIVERSITY: YIELD DYNAMICS OVER A FIVE-YEAR PERIOD. R.S. Chandran*, West Virginia University, Morgantown, WV (143)

ABSTRACT

Ten large-plot field experiments were conducted in West Virginia, USA over a period of five years (2010 to 2014) at three grower locations, to compare banded and broadcast applications of residual herbicides on corn yield and floral biodiversity levels. A pre-mixture of atrazine, metolachlor, and mesotrione was applied broadcast to provide 0.84, 2.24, and 0.224 kg ai/ha, respectively, or banded 38 cm wide over corn rows spaced 75 cm apart to provide the same herbicides at 0.42, 1.12 and 0.112 kg ai/ha, respectively, prior to weed emergence when the corn was 10 to 15 cm tall. In all years and locations, corn yields recorded were similar between banded and broadcast plots. Numerical differences amounted to 7.3 percent reduction in corn yield in plots that received banded application of herbicides compared to those that received broadcast application of the same, when averaged over years and locations (ranged from -2 to 15 percent). An analysis for Shannon's Diversity Index generated H values of 1.88 for banded plots compared to 1.49 in plots that received broadcast application of herbicides; according to Shannon's Index an *H* value > 1.5 is considered to be biologically-diverse. Simpson's Diversity Index analysis generated *1-D* values of 0.63 and 0.67 in banded and untreated plots, respectively, reflecting probability levels of two randomly-selected weeds being different species one month after herbicide application. Broadcast application resulted in total weed control, thus lacking a *1-D* value. Buildup of weed seed bank was a primary concern among growers. Herbicides may be band-applied only when expected weed populations are below a certain threshold, and mechanical methods along with cultural methods such as using cover crops may have to be implemented to reduce the buildup of weed seed bank. Banding herbicides may not only reduce use of herbicides such as atrazine but may also provide other benefits such as reduced soil erosion and nutrient runoff, habitat for beneficial insects and natural enemies, increased levels of carbon sequestration. Services provided by vascular plants to the ecosystem can be affected by reductions in floral diversity and strategies to restore such reductions in cornfields without affecting yields significantly may be worthwhile.

OCCURRENCE AND IMPACT OF SOYBEAN VEIN NECROSIS IN DELAWARE. N.M. Kleczewski*, W.J. Cissel, and J. Whalen, University of Delaware, Newark, DE (144)

ABSTRACT

Soybean Vein Necrosis disease (SVNd) is caused by Soybean Vein Necrosis Virus, a recently described viral pathogen of soybeans. The virus is a member of the Tospovirudae and is known to be vectored by soybean thrips in a persistent manner. Although SVNd is considered to be the most prevalent viral disease in soybeans, little is known of its epidemiology, impacts, and management options. We surveyed 50 full and double crop soybean fields in the 2015 season for appearance and severity of the disease. In addition, two small plot research trials were conducted to examine the impacts of thrips management and agronomic practices on viral expression, yield, and seed size. Trial 1 examined the effects of sequential foliar insecticide treatments, whereas Trial 2 examined ten soybean varieties and planting date effects. Survey results indicated that in 2015, 72 percent of Delaware soybean fields had detectable levels of disease by R5-R6, with 69 percent of full season and 93 percent of double crop fields affected. The within field severity ranged from 41 percent to 53 percent in full season vs double crop fields, respectively. Although insecticide treatments significantly reduced thrips numbers in Trial 1, there were no effects on SVNd symptomology. SVNd indices were greater in double crop vs full season beans (5.46 vs 2.43). In both systems, significant differences were detected for variety and SVNd index. In the full season trial, the varieties H42-13R2, 74B42R, and H37-14R2STS had significantly lower SNVd indices than other tested varieties. In the double crop system SVNd indices were significantly lower in variety 74B42. In both systems, SVNd indices were greatest in the varieties MA3516NRR2 and SG3644. Regression analyses indicated a significant negative linear relationship between SVNd index and 500 seed weight and yield. Survey results indicate that SVNd can be present at high levels in Delaware fields and may be greater in double crop systems. Our data are the first to document the effects of variety and planting date on SVNd and its associated effects on yield and seed size, and indicate that the virus may potentially impact yield to some degree. Based on our results, planting date and variety selection may be effective SVNd management practices.

VALIDATION OF PREDICTED MIRNAS IN *PHYTOPHTHORA SOJAE* AND *PHYTOPHTHORA INFESTANS*. C. Madison and M. Ospina-Giraldo*, Lafayette College, Easton, PA (5)

ABSTRACT

The post-transcriptional regulatory environment in oomycetes, especially in the economically important *Phytophthora infestans* and *P. sojae,* is very poorly understood. In particular, the expression of microRNAs (miRNAs), and their potential roles in infection, is a topic of increasing interest. In this study, we attempted to experimentally verify the existence of five miRNAs from both *P. infestans* and *P. sojae*, which had been predicted using *in silico* approaches. In addition, we investigated the possibility of these miRNAs playing a role in *Phytophthora* pathogenicity. The presence of two of the five miRNAs, namely psj-miR8788, predicted based on its stem-loop structure and sequence conservation between *P. sojae*, *P. ramorum*, and *P. infestans*, and PimiRNA3, predicted in *P. infestans*, was experimentally confirmed *in vivo* by amplification using specific stem-loop primers and subsequent cloning and sequencing. The predicted targets of psj-miR8788 include members of the amino acid/auxin permease family, which serve important amino acid transport functions, while the targets for PimiRNA3 include a double strand break repair protein, a histone arginine demethylase, an aspartyl-tRNA synthetase, an hsp70-like protein, and a mitogen-activated protein kinase. With such a large variety of predicted targets, it is difficult to determine whether these miRNAs play an important role in the infection cycle. However, in the future we hope to gather expression data for these two miRNAs in order to determine whether any changes in expression levels occur during infection, which could imply a pathogenic function.

EVALUATING RATE AND TIMING EFFECTS OF FACET L APPLICATIONS ON GRASS SPECIES IN THE GREENHOUSE. L. Vincent, W.J. Everman, and J. Copeland*, North Carolina State University, Raleigh, NC (148)

ABSTRACT

Due to limited options, grass weed management in grain sorghum (*Sorghum bicolor*) production is challenging. Facet L (quinclorac), released in 2013, is an option for control of grass weed species that can be applied preplant incorporated, preemergence, and postemergence. Previous research has provided that herbicide rate and timing can affect overall herbicide efficacy on target grass weed species. BASF label for Facet L provides that applications should be made on grass weed species no larger than 5.08 cm in height. Given the recent rise in acres of grain sorghum in North Carolina, growers should be aware of the importance of timeliness and rate for controlling the grassy weeds spectrum. Therefore, the objective of this study was to evaluate the response of six grassy weed species at three distinct growth stages with applications of Facet L at various rates.

Studies were conducted at the Method Greenhouse Facility in Raleigh, North Carolina in 2015 to evaluate the rate and timing effects of Facet L applications on six grass weed species commonly found in grain sorghum production in North Carolina. Grass weed species included large crabgrass (*Digitaria sanguinalis*), goosegrass (*Eleusine indica*), broadleaf signalgrass (*Urochloa platyphylla*), fall panicum (*Panicum dichotomiflorum*), texas millet (*Urochloa texana*), and crowfootgrass (*Dactyloctenium aegytium*). Facet L was applied at 290 and 420 g ai ha^{-1} with the addition of crop oil concentrate (COC) at 2.34 L ha^{-1} as well as an untreated check. Applications were made when species reached 2.54, 5.08, and 10.16 cm in height. Experiments were conducted using a factorial arrangement of treatments in a randomized complete block design with three factors being species, rate, and timing. All data were subjected to analysis of variance and means were separated using Fisher's Protected LSD at p=0.05.

Visual control 14 DAT of large crabgrass (92 percent) and fall panicum (87 percent) was significantly greater than compared to applications at 5.08 and 10.16 cm.

Crowfootgrass and goosegrass were not controlled by Facet L applications 14 DAT regardless of timing or rate. Height reductions 14 DAT at the 2.54 cm timing of large crabgrass (97 percent) and broadleaf signalgrass (96 percent) were significantly greater than later timings, 5.08 cm and 10.16 cm, ranging from 60-81 percent height reduction. Dry weight reductions for broadleaf signalgrass at all timings ranged from 92-99 percent. Dry weight reductions of large crabgrass with applications of Facet L at the 2.54 cm (99 percent) and 5.08 cm (96 percent) was significantly greater than the 10.16 cm (83 percent) timing. These data provide the importance of timely applications of Facet L for grassy weed species, specifically broadleaf signalgrass, large crabgrass, and fall panicum.

EVALUATION OF LATE SEASON FUNGICIDE APPLICATIONS ON FOLIAR DISEASE, YIELD, AND SILAGE IN CORN. A.A. Collins and N.M. Kleczewski*, University of Delaware, Newark, DE (149)

ABSTRACT

Weed control remains a major challenge for economically viable sorghum production in the Southeastern region of the United States because of sorghum sensitivity to weed competition during early growth stages. Field experiments were conducted in 2012, 2013, and 2014 to determine the effects of row spacing, sorghum population, and herbicide programs on Palmer amaranth, sicklepod and morningglory control as well as on sorghum growth and grain yield. Treatments included: three row spacing, 19, 38, and 76 cm; four sorghum populations, 99,000, 198,000, 297,000, and 396,000 plants ha^{-1}; and three herbicide programs, (1) a non-treated control, (2) a PRE application of prepackaged *S*â€‘metolachlor plus atrazine at 100 percent of the recommended rate, referred to as PRE, and (3) a PRE application of prepackaged *S*â€‘metolachlor plus atrazine at 75 percent of the recommended rate followed by early POST application of 2,4-D, referred to as PRE followed by POST (PRE fb POST). Palmer amaranth control for all locations benefited from the addition of a POST herbicide and also by increasing the sorghum population from 99,000 to at least 297,000 plants ha^{-1}. Palmer amaranth response to row spacing was variable across rating dates and years. Narrower row spacing or increased sorghum population did not affect Palmer amaranth density but caused significant dry biomass reduction by 33 percent with 19 and 38 cm compared to 76 cm, and by 43 percent with 297,000 or 396,000 compared to 99,000 plants ha^{-1}. Our results underscored the need for a POST application combined to sorghum population ≥ 297,000 plants ha^{-1} to consistently maintain ≥90 percent late season morningglory control. Light interception by the sorghum canopy was little or not affected by row spacing. However, sorghum population had large influence with canopy closure occurring one and a half weeks earlier for 297,000 or 396,000 plants ha^{-1} density compared to 99,000 plants ha^{-1}. Consistent grain yield increase by 18 percent on average was observed for 19 cm rows compared to 38 and 76 cm whereas sorghum plant populations used here had little or no effect. Overall, results from these experiments indicate that in the absence of POST application, narrow row spacing and sorghum populations of 297,000 plants ha^{-1} or more provide greater broadleaf weed control and biomass reduction.

EVALUATING THE EFFECT OF PRE-HARVEST SORGHUM DESICCATION ON SUBSEQUENT WINTER WHEAT YIELD. M.K. Bansal*, North Carolina State University, Raleigh, NC (151)

ABSTRACT

North Carolina sorghum production has renewed interest in recent years as regional grain demands increased which lead swine producer to offer a competitive sorghum grain price. Sorghum can be a good potential alternative for corn, soybean when planted in rotation with wheat. Sorghum has some advantages over corn and soybean. Sorghum can tolerate heat stress well compared to corn and also is less favored by deer compared to soybean. However, sorghum root hairs produce allelochemical 'sorgoleone' which has negative impact on following crop. Sorgoleone production is a continuous process as long as root hairs are functional.

Experiments were conducted in 2013-14 at Rocky Mount and 2014-15 at Kinston, North Carolina to investigate the effect of pre-harvest desiccation of different sorghum varieties on following wheat yield. Six sorghum varieties were planted in 2013-14 and two in 2014-15. Two weeks before harvesting, sorghum was either not desiccated or desiccated with glyphosate @ 1.37 lb ai/acre. In both years, there was no significant effect of sorghum desiccation on wheat yield. There was no significance difference between yields of different varieties of sorghum for desiccation treatment in both years.

EVALUATION OF THE IMPACTS OF ACID WHEY ON SOIL PH AND LIQUID MANURE PH. S.K. Gami*, Q.M. Ketterings, K. Czymmek, and G.S. Godwin, Cornell University, Ithaca, NY (154)

ABSTRACT

Acid whey (AW) is a by-product produced during the making of acid types of dairy products such as yogurt and cream cheese. Land application of AW is considered due to the nutrient content of acid whey and as a whey disposal mechanism. Our objective was to determine the effect of AW application, with or without mixing with liquid manure, on soil pH changes over time across representative soil types in New York. Two incubation studies were conducted. For study 1 (soil incubation), four AW rates (0, 10, 20 and 40 thousand gallons acre^{-1}) were applied to seven New York soils and incubated in the dark for 84 days at room temperature and 75 percent field capacity. For study 2 (whey and manure incubation), three liquid sources of manure and two sources of AW were collected. Acid whey and manure were placed in 1 liter jar at 1:0, 1:1, 1:2, 1:3, 1:4, 1:5, 1:8 and 0:1 ratios and incubated for 84 days. Both experiments were conducted in four replications and pH was measured 0, 1, 2, 4 and 7 days after mixing followed by weekly sampling until day 84. In study 1, the pH decrease measured after 1-2 days ranged from 0.14 to 1.44 pH units depending on application rate, original pH of the soil, and soil type. Soil pH increased to reach a maximum of 1.57 pH units above the untreated soil at 14 days after application. Soil pH decreased toward the end of the experiment in all soil types. At day 84, the pH was lower than the initial soil pH (0.12 to 1.24 pH units) for all soils except a sandy Barbour soil with a very low initial pH (pH 4.95). The decrease in pH at day 84 was proportional to the rate of whey applied ranging from 0.16 to 0.60 pH units with the highest AW application rates. The decrease was lowest for soils with an initial pH of 6.55 or higher. In study 2, the trend in pH measurement of all three sources of manure with acid whey at different time interval was similar; in general, the pH of the mixture increased with decrease in ratio of AW:manure. Ratios of 1:4 and 1:8 (AW:manure) increased from a pH of 6.3 to 6.4 on day 0 to 7.7 to 7.8 on day 84). Three additional studies are being conducted to assess the relationship between NH_3 and CO_2 dynamics over time and change in pH for seven different soil types. Field work is ongoing to also determine infiltration rates of AW and manure mixtures for various soil types.

MICROBIAL CO_2 RESPIRATION RESULTS RELATED TO SOIL HEALTH. W.J. Sciarappa*, S. Murphy, V. Quinn, and R. Barresi, Rutgers University, Freehold, NJ (155)

ABSTRACT

Fertilizer recommendations are typically based upon a chemical laboratory analysis and estimated crop needs over the growing season. The ability of farmland soils to biologically produce their own nutrients such as nitrogen through mineralization by soil microbes is not typically measured. This contribution to fertility from organic matter mineralization in the soil or from added compost amendments was measured by the Solvita® CO_2 respiration test. Four hundred standard soil samples were collected, processed and dried. Fungi and bacteria were then revived by adding water to cause a burst of respiratory carbon dioxide. The CO_2 was measured in two ways; either using a colored gel response chart or CO_2 measurement in ppm with a digital colorimeter. On a 0 to 5 scale, this biological nitrogen contribution ranges from 5 to over 50 lbs. nitrogen per acre which may be deducted from total crop needs. County extension staff incorporated this biological sampling method within a standard chemical soil test program. Results of this method in a two year pilot study in a diversity of agricultural systems show consistent and usable results over time and distinct differences among various cultural systems. Equipment and material costs were $13.62 per sample; not including time and labor costs. This low-cost, high-tech soil test has shown promise to provide a simple and quantitative means in surveying the current status of soil health or measure impacts over time from management practices such as tillage systems, cover cropping, compost amendments and chemical fertilizers. With this new assessment method, producers, advisers and extension agents add another evaluative parameter to their soil nutrient recommendations.

HERBICIDE AND NITROGEN APPLICATIONS IMPACT NITROUS OXIDE EMISSIONS. A.M. Knight*, W.J. Everman, S. Reberg-Horton, S. Hu, D.L. Jordan, and N. Creamer, North Carolina State University, Raleigh, NC (156)

ABSTRACT

Estimates of greenhouse gas emissions indicate that agriculture accounts for approximately 10 percent of these emissions. Agriculture is estimated to contribute largely to the output of one of the main greenhouse gases, nitrous oxide, which is suspected to be 59 percent of those emissions caused via agriculture. Additionally, this gas is thought to be contributing to climate change. While these large percentages are evidenced to be largely due to nitrogen application and tillage, little research has been conducted to determine if herbicides play a role in greenhouse gas emissions. An incubation study was conducted with treatments of the herbicides nicosulfuron, chlorimuron, flumioxazin, atrazine, glyphosate, glufosinate, pendimethalin, paraquat, mesotrione, isoxaflutole, s-metalochlor, 2,4-D, dicamba and a non-treated control. Additionally, high (22.5 kg ha-1) and low (135 kg ha-1) rates of urea ammonium nitrate were applied. Lab studies were conducted using a Tarboro Loamy Sand soil from Goldsboro, NC maintained at a water filled pore space of 70 percent. Gas flux was measured following treatment application at 12, 25, 48, 72, 120, and 168 hours for the greenhouse gas N_2O using gas chromatography. Results indicated that differences in N_2O emissions were present for both nitrogen rates, herbicides applied and the interaction. A field study was also conducted to quantify herbicide differences in the field following USDA-ARS GRACEnet Project Protocols. Gas measurements were taken at 24 hour increments for a week and, field results showed evidence of herbicide differences in N_2O emissions.

DEVELOPING A NEW P INDEX FOR NEW YORK WITH STAKEHOLDER INPUT. S. Crittenden*, S. Cela, Q.M. Ketterings, and K. Czymmek, Cornell University, Ithaca, NY (157)

ABSTRACT

Phosphorus indices (PIs) are tools designed to estimate relative risk of phosphorus (P) loss from agricultural fields based on P sources and their risk of transport to surface and ground water. These PIs attempt to encourage implementation of best management practices (BMPs) to reduce risk of P loss. The current New York PI (NY PI) was introduced in 2001. A survey of 36 certified nutrient management planners (CNMPs) in NY was conducted in 2014 to identify where they felt improvements to the NY PI might be focused. Planners thought that all existing source factors including soil test P, fertilizer P application (rate, timing, and method), and organic P application (rate, timing, and method) were important. Transport factors including soil drainage, flooding frequency, flow distance to nearest water body, and soil erosion rate were also important factors that should be included in the NY PI. Planners suggested that the various weighting factors in the NY PI should be re-evaluated and that best management practices such as manure incorporation, use of cover crops, setbacks and buffers to water bodies, and preferential manure applications to fields without connectivity should be incentivized. Similarly, it was suggested that high risk practices such as manure application to saturated or frozen soil, application to fields near water bodies, application to steeply sloped fields, manure spreading without incorporation, and high manure rates should be discouraged. Analyses of PI data from 33 327 NY farms were conducted to identify the main drivers of the current NY PI and to explore impacts of modified weighting factors. Flow distance to the nearest water body was the main driver of the transport factor and soil test P levels were mainly responsible for the source factor. Adjusted transport factors had only minor impacts on the PI. In contrast, increased weightings of manure application rate, timing, and method greatly impacted a field's NY PI. The survey results and the findings of the evaluation of the 33 327 NY fields, resulted in a proposed new NY PI. The new NY PI includes soil test P as an application cutoff, determines a field's inherent risk of loss, and then applies best management practices. The new NY PI produces a recommendation that falls in one of three categories: (1) no manure is allowed; (2) manure is allowed at rates that do not exceed P removal for the crop; (3) manure applications can be N-based.

FALL TRAFFIC TOLERANCE AND SPRING RECOVERY OF BERMUDAGRASS CULTIVARS IN THE TRANSITION ZONE. M. Goatley*, B. Kraemer, S.D. Askew, and W.L. Askew, Virginia Tech, Blacksburg, VA (158)

ABSTRACT

The density and aggressive creeping growth habit of bermudagrass (*Cynodon ssp.*) makes it a desirable sports turf surface and its use continues to expand in the transition zone of the United States. However, when combining possible constraints in winter hardiness, a winter dormancy period of at least 5 months in the transition zone, and intensive traffic of a fall sports season (primarily American football), there are likely limitations in long-term persistence and performance of some bermudagrass cultivars for transition zone athletic fields. The defunct 2007 National Turfgrass Evaluation Program bermudagrass trial in Blacksburg, VA was used to evaluate 31 bermudagrass cultivars under simulated fall football traffic conditions. A modified Brinkman traffic machine was used to simulate an equivalent of 3 football games per week for 10 consecutive weeks during the fall seasons of 2013 and 2014. The plots were periodically evaluated in the fall by both visual (estimated percent bare ground in the trafficked plots) and mechanical (both Normalized Difference Vegetation Index, NDVI and Ratio Vegetation Index, RVI) methods in order to quantify overall fall traffic tolerance, turf health, and plot density. The plots were also visually evaluated in the spring (estimated percent bermudagrass greening) and with NDVI/RVI measurements. There were significant cultivar main effects for fall visual ratings on percent bare ground and spring visual ratings for percent bermudagrass greening. Latitude 36, Tifway, Northbridge, Yukon, SWI-1057, and Premier Pro were in the highest statistical category for wear tolerance ratings (i.e., the lowest percent bare ground values) across the fall rating periods and Yukon, Latitude 36, Northbridge, and Patriot had the highest spring greening percentages across the spring ratings. RVI measurements generally correlated with the visual ratings and were deemed superior to NDVI measurements in this trial due to a more accurate bare ground assessment in the trafficked plots.

MESOTRIONE EFFICACY AND FATE IN ANNUAL BLUEGRASS AND KENTUCKY BLUEGRASS AS INFLUENCED BY GROWTH STAGE. J. Yu* and P. McCullough, University of Georgia, Griffin, GA (159)

ABSTRACT

Mesotrione provides PRE and early-POST control of annual bluegrass during Kentucky bluegrass establishment from seed, but applications do not effectively control multi-tiller plants. The physiological effects of growth stage on efficacy and the basis of mesotrione selectivity between species is not well understood. The objectives of this research were to evaluate mesotrione behavior in these species at three growth stages: pre-tiller (3 to 5 leaves), 1-tiller, and multi-tiller (5 to 7 tillers). In greenhouse experiments, single mesotrione application at 0.28 kg ai ha^{-1} injured pre-tiller, 1-tiller, and multi-tiller annual bluegrass 54, 33, and 11 percent at 4 weeks after initial treatment (WAIT), respectively. Sequential applications of mesotrione increased injury to pre-tiller and 1-tiller annual bluegrass by ~20 percent from a single treatment. Mesotrione reduced annual bluegrass tiller counts ~60 percent from the nontreated at the pre-tiller and 1-tiller stage. Sequential mesotrione applications caused ≤14 percent injury to multi-tiller annual bluegrass and Kentucky bluegrass at all growth stages, and did not reduce tillering from the nontreated. Annual bluegrass absorbed 34 percent more root-applied ^{14}C-mesotrione than Kentucky bluegrass in hydroponic culture, but relative differences (Bq g^{-1}) among growth stages were not detected. Annual and Kentucky bluegrass absorbed 31 percent and 35 percent of the applied radioactivity after foliar treatments, respectively. Differences in ^{14}C translocation and metabolism were not detected between species, and foliar uptake levels were similar across growth stages. However, multi-tiller plants metabolized ~2-times more ^{14}C-mesotrione than pre-tiller and 1-tiller plants. Overall, the selectivity of mesotrione for annual bluegrass control during Kentucky bluegrass establishment results from differential levels of root absorption. Mesotrione has limited efficacy for controlling multi-tiller annual bluegrass due to enhanced degradation compared to pre-tiller and 1-tiller plants.

EFFECTS OF INDAZIFLAM FORMULATION ON EARLY POSTEMERGENCE CONTROL OF *OXALIS STRICTA*. C. Marble*, A. Chandler, and M. Archer, University of Florida, Apopka, FL (160)

ABSTRACT

Due to lack of selective postemergence herbicides, container nursery growers must rely on preemergence herbicides and supplemental handweeding for weed control. Consequently, weed control is consistently ranked as one of the highest production costs in container production of ornamental plants. Woodsorrel species (*Oxalis* spp.) are consistently ranked among the most common and troublesome weeds in southeastern nurseries. Many preemergence herbicides offer control of woodsorrel, but it is common for small woodsorrel and other weeds to be missed or overlooked during handweeding which results in poor preemergence control and increased weed pressure. Over the last several years, several preemergence herbicides have been shown to have early postemergence control of small weeds, including dimethenamid-p for spotted spurge (*Chamaesyces maculata*), isoxaben for bittercress (*Cardamine* spp.), and indaziflam for yellow woodsorrel (*Oxalis stricta*). While dimethenamid-p and isoxaben are available in formulations labeled for use over-the-top of nursery plants, the indaziflam formulation previously tested is not. There has been interest among growers as to whether or not the new granular (G) formulation of indaziflam, which is labeled for over-the-top application, will provide similar control to the suspension concentration (SC) formulation which is not. In order to evaluate the efficacy of the granular formulation of indaziflam in comparison with the liquid formulation, yellow woodsorrel seed (approximately 50) were surface sown by hand to nursery containers filled with a pinebark:peat potting media (Fafard 52 growing mix, SunGro Horticulture, Agawam, MA) and standard amendments at two separate dates (Mar. 19 and Mar. 30, 2015) and grown to reach two different sizes: two to four leaf stage and six to eight leaf stage. On April 14, liquid (Marengo, OHP. Inc., Mainland, PA) and granular (Marengo G, OHP, Inc.) indaziflam formulations were applied at 12.6, 25.1, 50.2, and 100.4 kg ai/ha which would represent 0.125, 0.25, 0.5 and 1.0 times the highest labeled rate (×) of the liquid and 0.25, 0.5, 1.0, and 2.0× of the granular formulation. Glyphosate (RoundUp Pro, Monsanto, St. Louis, MI) was applied at a rate of 1.1 kg ai/ha and served as a chemical control. All herbicides were applied using a CO_2 backpack sprayer calibrated to deliver 936 l/ha. A non-treated control was also included. All plants were dry at the time of application and were not watered for four hours after application. The study was designed as a randomized complete block design with 10 single pot replications per treatment, per growth stage, and was repeated three times. Data collected included visual ratings at 1, 2, and 4 weeks after treatment (WAT) (not shown) and percent control which was based on the formula [((Nontreated – treated))/nontreated)*100)] using plant shoot fresh weight taken at 4 WAT. All data were subjected to ANOVA in SAS software (SAS, Cary, NC) and treatment means were separated using Fisher's Protected LSD test at the 0.05 significance level. Percent control data showed that indaziflam applied at 12.6 kg ai/ha (0.125×) provided similar postemergence to glyphosate for both growth stages in all three experiments. In general, the SC formulation outperformed the G formulation at the lowest rate (12.6 kg ai/ha) but G performed similarly to the SC formulation at the 50.2 kg ai/ha rate (labeled rate for G) and at the highest rate (100.4 kg ai/ha, 2x the labeled G rate). The labeled G rate of 50.2 kg ai/ha provided 85 to 100 percent control when applied to woodsorrel at the 2 to 4 leaf stage and 68 to 96 percent control when applied at the 6 to 8 leaf stage. Results from this trial indicate that the G formulation of indaziflam provides early postemergence control of yellow woodsorrel when applied at the highest labeled rate. While further work is needed to determine the early postemergence efficacy of the G formulation on other nursery weeds, the use of the G formulation of indaziflam could benefit nursery growers who have heavy infestations of yellow woodsorrel.

MITIGATING PRE HERBICIDE INJURY IN CONTAINER PRODUCTION OF *HYDRANGEA MACROPHYLLA*. J.C. Neal* and C.D. Harlow, North Carolina State University, Raleigh, NC (161)

ABSTRACT

Hydrangea macrophylla is a popular flowering shrub, and thus is produced in large numbers by container nurseries. The crop is relatively easily propagated and produced but is injured by most preemergence herbicides. Of the 25 preemergence herbicides labeled for use in container nursery crops, only 7 are labeled for use on *Hydrangea macrophylla* including Freehand, OH2, Snapshot TG, Pennant Magnum, Regalkade G, and Regal OO, and Surflan. However, grower experiences and research had demonstrated that each of these herbicides may injure *Hydrangea macrophylla* in container production. In 2015 we reported that spray-applied dimethenamid-p caused less injury than many common granular herbicides. In the current test we evaluated the hypothesis that irrigation scheduling may be utilized to mitigate spray-applied preemergence herbicide injury to hydrangea.

Hydrangea macrophylla 'Endless Summer' rooted cuttings were potted on August 25, 2015 to 4-L plastic pots using a pine bark + sand (9:1 v/v) substrate. Plants were placed in a shade house, then hand watered to settle the substrate. When foliage had dried, treatments were initiated. Dimethenamid-p EC and isoxaben SC + prodiamine 4L were applied at labeled rates to actively growing plants using a CO2 pressurized sprayer equipped with 2-8004 flat fan nozzles and calibrated to deliver 30 GPA . Treatments were applied 2 hours before irrigation, 15 minutes before irrigation, or during an overhead irrigation cycle. Applications made 2 hours before irrigation allowed sufficient time for spray treatments to dry on the foliage before irrigation. Applications 15 minutes before irrigation were not dry before irrigation was started. Irrigation was run long enough to thoroughly wet the foliage (about 10 minutes), then herbicides were applied while the irrigation was running. Total irrigation applied was about 0.5 inch. The treatments were assigned in a randomized complete block design with 4 replications and 3 plants per plot. Treatment effects were evaluated in several manners. Visual injury ratings were conducted periodically starting 1 week after treatment using two scales. First, an injury rating using a 0 to 10 scale where 0 = no injury (equivalent to non-treated plants) and 10 = dead plants. Secondly, each plants was evaluated on a "plant quality" scale where 0 = dead plants and 10 = the best plant quality and size in the study. Ten days after treatment, the number growing points with necrotic symptoms were counted and expressed as a percent of the total number of growing points present on each plant. Plant heights and widths were measured and used to calculate estimated plant canopy volumes. Plant fresh weights were measured at the termination of the study.

Applications made to dry foliage before irrigation resulted in significant injury, but applying herbicides during an irrigation cycle significantly reduced tip necrosis and improved hydrangea growth and quality compared to applications made before irrigation. Isoxaben + prodiamine treatments were more injurious than dimethenamid-p within each irrigation regimen. Measured 10 days after treatment, up to 68 percent of the growing points were necrotic on plants treated with isoxaben + prodiamine before irrigation, compared to less than 20 percent on plants treated with dimethenamid-p 2 hours before irrigation. When applied during an irrigation cycle, dimethenamid-p caused no injury but isoxaben + prodiamine plants had an average of 11 percent tip necrosis. By 55 days after treatment, plants treated during an irrigation cycle were similar to the non-treated plants in height, width and fresh weight

POWDERY MILDEW MANAGEMENT WITH BIOPESTICIDES IN GREENHOUSE GROWN ZUCCHINI. B.C. Eshenaur* and E.M. Lamb, New York State Integrated Pest Management, Rochester, NY (162)

ABSTRACT

Many greenhouse growers who produce herbs and vegetables in their crop mix are increasingly seeking alternative plant disease control products. Trials were established at the at the Barton Greenhouses of Cornell's New York State Agriculture Experiment Station in Geneva, NY. We evaluated the efficacy of selected OMRI certified biopesticides on powdery mildew. Test plants used in these studies were zucchini cultivars with know susceptibility to powdery mildew, 'Raven' and 'Zephyr'. In early trials no significant cultivar effects were observed. In later trials only the cultivar 'Zephyr' was used. In a later study the zinnia cultivar 'Envy' was also used for a powdery mildew trial. The biopesticides evaluated in these studies included: Actinovate (*Streptomyces lydicus*) Cease *(Bacillius subtilis),* MilStop (Potassium bicarbonate), Regalia (Extract of *Reynoutria sachalinensis)* SuffOil-X (Formulated mineral oil) and Sonota (*Bacillus pumilus*) as well as water as a control treatment. Randomized block designs were used with at least five replicates in each treatment in each trial. Study plants were inoculated and became diseased with powdery mildew. Spray were applied weekly. Plants were rated visually for the percentage of foliage that was symptomatic. Some of the products used showed statistically significant disease reduction in powdery mildew over the control treatment; the products that performed the best included: Milstop, Regalia and SuffOil-X.

NOVEL PCR PRIMERS FOR DETECTING BACTERIAL ROT PATHOGENS OF ONION. J.E. Asselin*, J.M. Bonasera, and S.V. Beer, Cornell University, Ithaca, NY (163)

ABSTRACT

Bacterial rots of onion are a serious problem in New York State. Growers sometimes sustain losses up to 40 percent. Numerous bacteria have been implicated as bulb-rotting pathogens, with symptoms ranging from mild discoloration of a single internal scale to maceration of the entire bulb. Mixed infections of a single bulb are common, and bulb symptoms are not typically distinct enough to implicate which bacteria are responsible for symptom development. In New York State, three groups of onion-pathogenic bacteria are problematic: bacteria of the *Burkholderia cepacia* complex and *Burkholderia gladioli*, *Pantoea ananatis*, and strains of the *Enterobacter cloacae* complex. To facilitate detection of specific bacterial pathogens of onion, we leveraged available genome sequences to design specific PCR primer pairs to amplify DNA from each of these groups of bacteria. Primers were validated against dozens of target strains (and closely related non-target organisms) collected from onions and other sources in New York and elsewhere. The primers detected bacteria in pure and mixed cultures by visualization of single, discrete bands in agarose gels. The novel primer pairs have proven highly useful in identifying pathogens isolated from symptomatic bulbs submitted to the laboratory.

PREVALENCE OF MALE FLOWERS IN SPRING SEEDED ZUCCHINI SQUASH CULTIVARS.

M.L. Infante-Casella*, Rutgers NJAES Cooperative Extension, Clarksboro, NJ (164)

ABSTRACT

Zucchini squash yields are often hindered by the fact that female flowers will open 5-7 days before male flowers. Since male flowers are not yet opened, pollen is not able to be transported by insect pollinators to the female flowers. Hence, seeds within the developing fruit are sterile. Seeds that are fertilized produce hormones that promote fruit enlargement and uniform fruit formation. Developing fruit with sterile seeds will not form properly and may abort. Therefore, the first week of harvest may result in culled fruit that is discarded. Since harvest of commercial zucchini squash fields only lasts 4-5 weeks, losing the first week marketable fruit can dramatically reduce yield and profit. A preliminary study was conducted in spring of 2010 to evaluate the timing and prevalence of male flowers on green zucchini squash cultivars. The study contained a randomized complete block design with 3 replications and 6 plants per plot. Eleven cultivars were seeded by hand with 2 seeds per hole, on April 21. Field culture consisted of 6 in. raised beds with black plastic mulch and drip irrigation. Rows were 60 in. apart and plants 30 in. apart within the row. After germination and formation of the first true leaf, plants were thinned to one healthy plant per hole. The field was observed daily beginning on June 2 to record the opening of the first male flowers. This did not occur until June 8. At this initial date, 6 cultivars having the highest numbers of open male flowers were: ‘Zucchini Select’, ‘Senator’, ‘Leopard’, ‘Judgement III’, ‘Spineless Beauty’ and ‘Tigress’. After the initial start date, newly opened male flowers were recorded and removed from the plants daily between 8:00AM and 10:00AM. Cultivars showed no significant difference in the number of male flowers opened over a 10 day period.

IR-4 PROJECT TRANSITIONS IN THE NORTHEAST REGION AND BEYOND.

J. Baron*, M. Ross, and D. Kunkel, IR-4 Project, Princeton, NJ (165)

ABSTRACT

The IR-4 Project was created as a partnership between the federal government, the states via the Land Grant Universities and State Agricultural Experiment Stations, the crop protection industry and the growers of fruits, vegetables, nuts, herbs, ornamentals and other specialty crops. In recent times, fiscal pressures within certain states have changed the dynamics of support of agriculture research. Cornell University, the longtime host institution for the IR-4 Northeast Region, now requires that all grants must contribute at least 18 percent of the grant funds to help them cover facilities and administration costs. The federal authorization language for the IR-4 Project funds does not allow for any indirect costs. Because of the federal limitation, Cornell University is phasing out their involvement with IR-4 Project. The impact of these changes in IR-4 Northeast Region research will be presented.

THE IR-4 PROJECT: UPDATE ON WEED CONTROL PROJECTS (FOOD USES). M. Arsenovic*, J. Baron, D. Kunkel, and R. Batts, IR-4 Project, Princeton, NJ (166)

ABSTRACT

The IR-4 Specialty Crop Program is a publicly funded program that develops and submits regulatory data for the registration of pest control products on specialty crops. IR-4 has a long history of providing herbicide registrations for specialty or minor crop growers. In 2015, IR-4 data was used to support a number of new herbicide registrations. These included new registrations for sethoxydim, S-metolachor for lettuce and other crops, and for halosufuron-methyl. EPA issued Notices of Filling for use of NAA, penoxsulam, fluazifop-p-butyl, clethodim, carfentrazone-ethyl and fomesafen. Herbicide submissions were made to the EPA in 2015 for the use of clomazone, fomesafen and flumioxazin in many crops. In addition, IR-4 submitted final reports to Registrant for label change for the use of sulfentrazone in mint, halosulfuron in cucumber and S-metoalchlor in Swiss chard. The 2016 IR-4 research plan will include about 15 percent herbicide projects. The IR-4 Project will remain a responsive and efficient organization that supports US stakeholders by facilitating U.S. regulatory approvals for crop protection tools. IR-4 will continue with a research focus on low risk pesticide registrations that support integrated pest management systems. There will be more emphasis on efficacy and crop safety data to determine products that can manage hard to control pests and mitigate invasive species. There will be continued support for products that can be used for organic growers as well as cutting edge biotechnology products. Other areas expected to have greater importance will be in the area of international harmonization, to assist in exports of U.S. grown specialty crops and removing pesticides as a technical trade barrier.

BROCCOLI VARIETY TRIALS IN SE PENNSYLVANIA. T. Elkner* and S. Bogash, Penn State Extension, Lancaster, PA (167)

ABSTRACT

Twenty five varieties of broccoli were evaluated in spring and fall plantings in 2014 and 24 varieties were evaluated in 2015 in SE Pennsylvania. 'Imperial' was the standard. Both spring trials and the fall 2014 trial were in bare soil; the fall 2015 planting was done using plasticulture. Marketable head number was higher in both years in the fall plantings but overall was lower in 2015 than 2014. Marketable head number for

Imperial was higher than 9 and 6 other varieties in the spring and fall of 2014, respectively, while it was only higher than 3 and 4 varieties in the spring and fall, respectively, in 2015. Mean marketable head weight for Imperial was not different than most varieties in the spring for both years while it was less than 8 varieties and greater than 3 in the fall of 2014. In the fall of 2015 there was no difference in mean head weight of all varieties. Broccoli growers in SE PA have many varieties available for production with acceptable yield and quality.

SINGLE VS SPLIT APPLICATIONS OF POST-EMERGENT HERBICIDES FOR SPREADING DOGBANE (*APOCYNUM ANDROSAEMIFOLIUM*) CONTROL IN WILD BLUEBERRY FIELDS. J.L. Cote* and D.E. Yarborough, University of Maine, Orono, ME (6)

ABSTRACT

Spreading dogbane (*Apocynum androsaemifolium*) is a major weed pest in wild blueberry fields, and is difficult to control with many of the industry's currently registered herbicides. In spring 2015 we initiated a trial at the University of Maine's Blueberry Hill Experiment Station Farm to examine the effect of two newer herbicides, mesotrione (Callisto) and rimsulfuron (Matrix), on dogbane control. Young dogbane was sprayed post-emergence to four 1 x 2 m plots either once at the full rate (mesotrione 6 oz/a and rimsulfuron 4 oz/a) or in a split application at the half rate (3 oz/a and 2 oz/a applied two times, respectively), both alone and as tank mixes, for a total of six herbicide treatments. In June and July wild blueberry and dogbane cover and phytotoxicity were compared to each other, to an untreated check and to the Blueberry Hill Farm's 2015 spray regime which was a 5/13 pre-emergence application of hexazinone 2 lb/a, terbacil 2 lb/a and diuron 1.6 qt/a, and a 5/27 post-emergence application of mesotrione 6 oz/a and clethodim 6 oz/a. Control of other broadleaf weeds and grasses were also assessed. It should be noted that at the June assessment, the split application treatments were assessed just prior to the second herbicide application. Initially in June, all treatments except rimsulfuron at 4 oz/a and the split 2 oz/a treatments had significantly more dogbane injury than the check, but no treatment significantly reduced dogbane cover. By July, dogbane injury approached 90 percent in the mesotrione split (3 oz/a) and mesotrione 3 oz/a + rimsulfuron 2 oz/a split applications and dogbane cover was reduced to less than 10 percent. Dogbane cover was significantly lower in all treatments compared to Blueberry Hill Farm's treatment, but only the split mesotrione treatment and the split mesotrione + rimsulfuron treatment were significantly lower than the check. The mesotrione and mesotrione + rimsulfuron tank mixes resulted in higher dogbane injury and lower cover overall, which suggested that mesotrione is more effective in controlling this weed. The lack of total control was due to new dogbane emerging after the herbicide applications; new dogbane shoots were observed at the second application of the split treatments and at the July assessment. The rimsulfuron 4 oz/a and mesotrione + rimsulfuron split treatments resulted in slightly lower blueberry cover, but phytotoxicity (chlorosis) was 20 percent or lower at both evaluations and so was considered acceptable. The split rimsulfuron and split mesotrione + rimsulfuron treatments resulted in the highest injury by July. Other weed cover in the plots was very low, below 12 percent overall, and there were no significant differences among treatments at either evaluation. In conclusion, the split mesotrione treatment was the most effective in controlling spreading dogbane; it resulted in the lowest dogbane cover and greatest dogbane injury along with the split mesotrione + rimsulfuron treatment, but considering the two were equal in cover and within 2 percent of each other in phytotoxicity, the addition of rimsulfuron did not improve the control and therefore is unnecessary.

EFFECTS OF ABSCISIC ACID (ABA) AND GIBBERELLIC ACID (GA3) APPLICATIONS ON CLUSTER LOOSENING IN TIGHT CLUSTER VARIETY *CHARDONNAY* IN SOUTH JERSEY. H. Gohil* and D. Ward, Rutgers University, Clayton, NJ (169)

ABSTRACT

A trial was conducted to study the efficacy of Abscisic Acid (ABA) and Gibberellic Acid (GA_3) on cluster loosening in *Vitis vinifera* cv. 'Chardonnay' at Rutgers Agricultural Research and Extension Center in southern New Jersey. A randomized complete block experimental design was employed with five blocks of six treatments. Treatments comprised of un-treated control, 100 ppm ABA applied once (80-90 percent bloom), 100 ppm ABA applied twice (80-90 percent bloom and 3 d later), 200 ppm ABA applied once (80-90 percent bloom), 200 ppm ABA applied twice (80-90 percent bloom and 3 d later) and 4 percent GA_3 applied once (pre-bloom). At harvest cluster looseness of whole cluster and the number of berries per mm, the length and average berry weight of second true shoulder were measured. The untreated control had the tightest while GA_3-treated vines had the loosest clusters. Rate and number of ABA applications loosened clusters at the varying levels. However, neither of the PGRs (Plant Growth Regulators) affected length or the number of berries per mm of the second true shoulder. Average berry weight was higher in ABA-treated berries applied twice at the higher rate. There were several small but fully matured berries in clusters of GA_3-treated vines which could have increased the cluster looseness. PGRs did not cause any phytotoxicity to treated vines.

EVALUATION OF SYSTEMIC INSECTICIDE AND FUNGICIDE FOR PROTECTION OF SYCAMORE FROM POLYPHAGOUS SHOT HOLE BORER / FUSARIUM DIEBACK. D.M. Grosman* and A. Eskelan, Arborjet Inc, Woburn, MA (170)

ABSTRACT

The polyphagous shot hole borer (PSHB), *Euwallacea sp.,* and its associated fungi has recently caused dieback and mortality of California Sycamore in southern California. Direct control option for PSHB were limited to frequent bark sprays. Recently, a systemic insecticide, emamectin benzoate alone or combined with a systemic fungicide, propiconazole has been shown to be effective in protecting pines against several pine bark beetles and their associated fungi. Two separate trials were established in 2013 and to evaluate efficacy of recommended rates of emamectin benzoate (TREE-age™) alone or combined with propiconazole (Propizol™) for protection or therapeutic treatment of individual California sycamore trees for PSHB/Fusarium Dieback. Treatments containing emamectin benzoate were effective in reducing the number of new attacks, level of sap flow on the bark surface, and tree mortality compared to untreated checks. Emamectin benzoate alone can protect trees for at least one full year. The combination of insecticide + fungicide, propiconazole, provides extended protection against new attacks and fungal infection through the second year.

EFFECTIVENESS OF TRICLOPYR-AMINE BASAL BARK TREATMENT COMPARED WITH STANDARD TRICLOPYR STEM TREATMENTS FOR EUROPEAN ALDER CONTROL. E. Weaver* and A. Gover, Penn State University, University Park, PA (171)

ABSTRACT

European alder (*Alnus glutinosa*) is an exotic, nitrogen-fixing tree that grows in near-aquatic habitats such as riparian corridors and lakeshores. Chemical treatments that prevent resprouting and the need for retreatment are efficient means of control, but due to the sensitive nature of aquatic systems, use of aquatic-labeled herbicides would significantly reduce non-target risk. Compared to foliar treatment, stem treatment should reduce herbicide contact with water, and could also be used to treat larger, taller stems than a foliar application. Standard stem treatment in these settings would feature glyphosate or triclopyr-amine administered via hack-and-squirt technique. An easier technique is the basal bark treatment, which is applied to intact bark. Triclopyr-ester is the most common basal bark treatment, but is contraindicated for near-aquatic and aquatic use due to label restrictions. The purpose of this experiment was to compare the standard triclopyr-amine hack-and-squirt and triclopyr-ester basal bark treatments with three triclopyr-amine basal bark treatments of differing concentrations, to determine if a triclopyr-amine basal bark treatment could effectively control the European alder. The experiment was established August 18 and 26, 2015 on an upland site at Bald Eagle State Park in Howard, PA, using five treatments plus a control, with ten trees per treatment, arranged as a completely randomized design. The standard treatments were triclopyr-amine applied at 0.14 g ae/cm diameter breast-height (DBH, measured at 1.37 m height) to fresh hatchet cuts in the stem; or triclopyr-ester applied as basal bark treatment at 0.071 g ae/cm DBH. The experimental treatments consisted of triclopyr-amine applied at 0.071, 0.11, or 0.14 g ae/cm DBH, diluted in a 1:1 mix of water and basal oil with emulsifer that constituted 75, 62, and 50 percent of the mix by volume, respectively, with increasing triclopyr rate. Each treatment was applied using a syringe after measuring each stem to the nearest mm. The treatment effectiveness was assessed by visually estimating percent canopy reduction on September 19, 2015. The ratings did not set untreated plants to an arbitrary zero, hence the untreated stems averaged 12 percent canopy reduction due to seasonal leaf drop. Data were subjected to analysis of variance, and means compared using Fisher's Protected LSD. Treatment effect was highly significant ($p=<0.0001$). Canopy reduction was greatest with the hack-and-squirt and basal bark (98 and 83 percent, respectively), and were not significantly different from each other. Triclopyr-amine basal bark treatments were rated at 45, 70, and 64 percent canopy reduction at 0.071, 0.11, and 0.14 g ae/cm DBH, respectively.

All treatments were significantly different from the untreated control. These preliminary results are very encouraging, and provide hope that an aquatic-compatible basal treatment may be viable. Final defoliation and mortality ratings will be conducted in summer 2016.

DETECTION AND PHYLOGENETIC RELATIONSHIPS OF *PUCCINIA EMACULATA* AND *UROMYCES GRAMINICOLA* (PUCCINIALES) ON SWITCHGRASS IN NEW YORK USING RDNA SEQUENCE INFORMATION. S.C. Kenaley and G.C. Bergstrom*, Cornell University, Ithaca, NY (172)

ABSTRACT

The diversity of rust fungi (Pucciniales) inciting disease on switchgrass (*Panicum virgatum*) grown in bioenergy systems in North America remains unclear. Therefore, the species number and phylogenetic relationships of rust isolates affecting switchgrass were examined in 2011–2013 at two sites in New York as well sites in Alabama, Iowa, Nebraska, Pennsylvania, South Dakota, and West Virginia using ribosomal RNA gene data. Uredinia and teliospore morphology were also utilized to delimit taxa. Maximum likelihood, maximum parsimony, and Bayesian analyses demonstrated two monophyletic clades. Clade I consisted of *Puccinia emaculata* and included the majority of isolates across sites, whereas, Clade II included multiple isolates from IA, NE, and SD. Uredinia among all isolates were morphologically indistinguishable. Teliospores for isolates assigned to Clade I-*P. emaculata* were two-celled, brown to chestnut-brown, and oblong to ellipsoid averaging 35.4 ±0.6 x 18.0 ±0.2 µm. Clade II isolates also possessed ellipsoid, brown to chestnut-brown teliospores; however, spores were one-celled and significantly shorter in length (26.1 ±0.4 x 17.5 ±0.3 µm) compared to *P. emaculata* and hence, diagnosed to *Uromyces graminicola*. Studies to examine the inter- and intraspecific genetic diversity as well as the population biology and taxonomy of both rust fungi are forthcoming.

PASTURE HERBICIDE STEWARDSHIP :OUR EFFORTS TO REDUCE OFF-TARGET DAMAGE TO SENSITIVE, HIGH VALUE CROPS. N. Rhodes* and D. McIntosh, University of Tennessee, Knoxville, TN (173)

ABSTRACT

Off-target movement of pasture herbicides to high value crops continues to be an issue. Damage can result in lost productivity for growers, expensive fines and/or lawsuits, and negative publicity for the industry. We began a comprehensive educational program in 2011 that stresses the importance of proper stewardship with the use of pasture herbicides. The program was created to reduce the occurrence and impact of off-target damage to sensitive, high value crops; and to make available tools to help with the diagnosis of suspected cases of off-target damage. Funding was obtained via grants from Philip Morris International, Altria Client Services, Dow AgroSciences and DuPont Crop Protection. Four crops (tobacco, cotton, tomato and grape) and five herbicides (2,4-D, dicamba, aminopyralid, aminocyclopyrachlor and picloram) were selected for the development of educational materials and diagnostic tools. These include still images, time lapse videos and fact sheets that were created and made available when our initial website, herbicidestewardship.utk.edu, was launched in 2014. In 2015 the website was completely redesigned in an effort to make it more attractive and user friendly. Additionally, a shorter, less cumbersome address, herbicidestewardship.com, was created. Use of the website has steadily increased since its inception. The website has been visited over 8000 times since it was launched, and approximately half of these visits were in 2015 alone. Visits came from The United States, China, Japan, Germany, Canada, The United Kingdom, India and Brazil.

THE EFFECTS OF ROADSIDE HABITAT ON INSECT TRAFFIC MORTALITY. W.A. Keilsohn* and D. Tallamy, University of Delaware, Newark, DE (174)

ABSTRACT

Paved roadways span over 4.1 million miles across the continental United States (U.S. Highway Admin. 2013). They are frequently bordered by both natural and restored habitats, and could provide important real estate for conservation. Because animals are frequently killed by auto traffic, there is concern that roadside habitats may be ecological traps, luring animals to their death rather than helping to build their populations. An alternative untested hypothesis is that good roadside habitat actually decreases animal mortality because it reduces the need for animals to leave the habitat. We tested this hypothesis in three roadside habitats: Woods, Meadows, and Lawns. We further confined our study only to roads with relatively high speed and heavy traffic volume. We divided our study sites into areas with and without a median strip and were careful to select areas where the median strip contained the same habitat as the edges of the road. We established thirty sites, five for each type of habitat both with and without a median, and at each site we surveyed a two hundred meter transect on each side of the road. We measured all animal mortality along each transect but our focus was on insects. We also counted all flowers within two meter of the edge of the road and collected all dead specimens under ten centimeters in size. We found that wooded areas along roads had significantly lower insect mortality when compared to lawn or meadow areas.

PROTEIN MOONLIGHTING IN PLANT DISEASE: CHARACTERIZATION OF THE DUAL FUNCTION OF A BACTERIAL TYPE III SECRETED VIRULENCE EFFECTOR AND CHAPERONE AS A TOXIN-ANTITOXIN SYSTEM. T. Shidore*, Connecticut Agricultural Experiment Station, New Haven, CT (175)

ABSTRACT

Toxin-Antitoxin (TA) systems are bacterial self-killing systems comprised of an antibacterial protein toxin and a neutralizing antitoxin. Found on bacterial mobile genetic elements as well as on bacterial chromosomes, TA systems are thought to be involved in variety of mechanisms like plasmid maintenance, stress response, antibiotic resistance etc. AvrRxo1 is a type III secreted effector of *Xanthomonas oryzae* pv. *oryzicola* (*Xoc*), the causal agent of bacterial leaf streak of rice, with homologs in other plant pathogens of *Xanthomonas*, *Burkholderia*, and *Acidovorax*. AvrRxo1 triggers a type III secretion-dependent hypersensitive resistance response (HR) in plants expressing the resistance protein Rxo1. *avrRxo1* homologs are all encoded upstream of a short gene coding for AvrRxo1 binding partner Arc1. The recently solved structure of AvrRxo1 revealed its similarity to a *Streptococcus pyogenes* TA toxin belonging to the zeta-epsilon family of TA systems. In this study, a standard validation assay is used to demonstrate that AvrRxo1:Arc1 is a functional TA system. A putative distant homolog of AvrRxo1 found in a soil myxobacterium *Cystobacter fuscus* demonstrated AvrRxo1-like plant toxicity and Rxo1-triggering activity, but did not share the bacteriostatic effect of AvrRxo1. These findings demonstrate that AvrRxo1:Arc1 is an example of a TA system with dual function as a Type III secretion-dependent avirulence gene.

ORGANIC PRODUCTION, BENEFICIAL ORGANISMS, AND BIOLOGICAL CONTROL

ORGANIC VERSUS CONVENTIONAL BEEKEEPING PHILOSOPHIES AND THEIR INFLUENCE ON HONEY BEE COLONY HEALTH.

R. Underwood*, D. vanEngelsdorp, B. Traver, and K. Nichols, University of Maryland, College Park, MD (176)

ABSTRACT

Honey bee colony health is jeopardized by a multitude of stressors, including lack of forage, agricultural and in-hive chemical use, parasites, pests, and viruses. Management practices used by beekeepers may aggravate or mitigate these stressors. Conventional management practices include the use of comb foundation, sugar syrup feed, antibiotics, and in-hive chemicals for pest control. Alternatively, organic beekeeping practices avoid the use of comb foundation, non-organic in-hive chemicals, and non-bee collected feed when possible. This study is a side-by-side comparison of honey bee health in colonies kept using these two management practices. Currently in the first year of a two-year project, we have measured comb production, varroa mite population growth, honey production, and queen supersedure rates and swarming frequency as measures of colony health. Thus, preliminary results will be shared.

DO MANAGEMENT PRACTICES IMPACT *NOSEMA CERANAE* LEVELS IN HONEY BEE COLONIES?

B.E. Traver*, R. Underwood, K. Nichols, and D. vanEngelsdorp, Penn State Schuylkill, Schuylkill Haven, PA (177)

ABSTRACT

With increased colony losses seen nationwide, many beekeepers wonder if there are ways in which they can reduce their own losses. There has been a movement in the "backyard beekeepers" towards using less pesticides when managing honey bees. Often, honey bees are exposed to several pesticides by the beekeeper as treatments for the various parasites, pathogens, and pests, especially to control the varroa mite and nosema disease. Beekeepers have adapted organic methods for keeping bees in the hope that the bees will have one less stressor and will be healthier. In this study, we are examining whether bees managed conventionally versus bees managed organically have different levels of *Nosema ceranae* or *Nosema apis* in the colonies. Data presented will be for the first year of the study.

INTEGRATION OF PELLETIZED POULTRY LITTER INTO ROLLED COVER CROPS FOR ORGANIC CORN PRODUCTION. G.M. Zinati*, J. Moyer, R. Atwell, S. Reberg-Horton, and S. Mirsky, Rodale Institute, Kutztown, PA (178)

ABSTRACT

In the absence of herbicides, organic grain producers rely on mechanical cultivation to manage weeds. Frequent soil cultivation weakens soil health; brings buried weed seeds to the soil surface, and adds to cost of production. The development of the roller-crimper has provided an alternative means for terminating cover crops and enables no-till planting of cash crops into rolled cover crop mulches that serve as a primary weed suppression tool. No-till system is gaining popularity among organic grain growers. However, this system creates challenges for livestock farmers who incorporate their operations' manure wastes into soil, as valuable fertility amendments, prior to traditional tilled systems. To address these challenges, a multi-institutional, multi-disciplinary team (Rodale Institute, USDA-ARS- Beltsville, and North Carolina State University) conducted on-station and on-farm trials to evaluate and demonstrate the impact of different starter fertilizer sources on yield and weed competition in organic no-till corn production. A cereal rye (*Secale cereale* L.) and hairy vetch (*Vicia villosa* Roth) cover crop (CC) mixture (60:40) was grown in all treated plots in the fall of 2012 and 2013, rolled and crimped at corn planting in spring 2013 and 2014. On-station trial included four treatments: CC (no fertility added), subsurface banding of pelleted poultry litter (CC+PL) at 12 kg N ha^{-1}, broadcast pelleted manure (BC) at 70 kg N ha^{-1} and Nature Safe pelleted chicken manure (NS) at 80 kg N ha^{-1}. At the on-farm trial the two treatments CC and CC+PL at 10 kg N per ha were selected and compared to the grower's standard practice (tilled in cover crop with 24.6 Mg/ha dairy manure).

Research was conducted at Beltsville, MD, Kinston, NC, and Salisbury, NC, from 2012-2014. Cover crop biomass in excess of 9,000 kg ha^{-1} and excellent weed suppression was observed at four of the six environments. In a combined analysis of four environments, corn grain yield was similar among the BC and NS fertility treatments, and was significantly lower with the CC+PL and CC treatments. Weed competition only affected corn grain yield at the lowest fertility environment. In this environment, the high BC fertility treatment had the lowest weed coverage, lowest weed biomass, and highest yield. In a high fertility environment excluded from the combined grain yield analysis, corn grain yield was similar across all fertility treatments, indicating that at a high fertility environment when high cover crop biomass levels are achieved, additional N fertility may not be necessary to maximize corn grain yield.

The on-farm trial was conducted at grower's field in Kutztown, PA. Mean cover crop biomass was 13,272 kg/ha. Although corn plant biomass was 21.7 percent greater in the grower's treatment than that in the CC + PL, corn grain yield in the grower's treatment was similar to that in the CC+PL treatment but significantly greater than that in the CC treatment. Weed pressure in the CC+PL was 29 percent lower than that in the grower's treatment and required no tillage during the growing season when compared to four cultivations in grower's treatment. The results from the demonstration trial indicate that the addition of banded subsurface pelleted poultry litter into rolled-crimped cover crop mulch boosts corn yield, reduces weed pressure, and eliminates tillage between rows during corn growth, leading to reduction in energy and labor costs. Results from the multidisciplinary studies indicate that starter fertilizer materials are often necessary to maximize corn grain yield in organic rotational no-till corn production and that producers have flexibility in selecting fertilizer materials and application methods in high fertility environments. At low fertility environments, considerable N provision is critical to ensure crop competitiveness with weeds and high corn grain yield.

COVER CROPS AFFECT WEED ABUNDANCE BUT NOT WEED COMMUNITY ASSEMBLY. B. Baraibar*, M. Hunter, M. Schipanski, and D.A. Mortensen, Penn State University, State College, PA (179)

ABSTRACT

Cover crops are increasingly being used to enhance ecosystem services like nutrient capture, erosion reduction and weed suppression. The contribution of cover crops to weed management can be particularly important in organic systems where synthetic herbicides cannot be used. Organic farming systems of the northeastern United States enjoy the highest levels of adoption of cover cropping in the country. In the region, cover crops are planted in the summer-fall period following corn and winter cereal grain harvest and grow through planting of the following cash grain crop. Understanding the ability of different cover crop species and mixtures to suppress weeds can inform farmers' decisions on what cover crops to use and the degree of weed suppression they can expect from them. In this experiment, we measured weed biomass in the fall and spring as an indicator of weed suppression by cover crops in the post-winter cereal grain window.

We tested 11 cover crops (six monocultures and five mixtures) in an organic systems experiment in central Pennsylvania, during three cropping seasons (fall 2012-spring 2015). Cover crop species included two legumes (red clover [*Trifolium pratense*] and Austrian winter pea [*Pisum sativum*]), two grasses (cereal rye [*Secale cereale*] and oats [*Avena sativa*]), two brassicas (tillage radish [*Raphanus sativus*] and canola [*Brassica napus*]) and combinations of these species in mixtures ranging from 3 to 6 species. We also included a 7-species mixture that included additional species and a no cover crop control.

Weed biomass in the fall differed greatly across the three years, which we largely attribute to three different cover crop planting dates. Planting cover crops in early August resulted in higher weed biomass compared to sowing in mid and late August across all cover crop treatments. Interestingly, summer annual weeds accounted for most of the fall weed biomass in all years and especially in the early planting year. Within each year, there were significant differences in fall weed biomass across cover crop treatments. Rye and oats monocultures and the 3-, 4- and 6-species mixtures were the most suppressive cover crops. In spring, weed abundances were low in all treatments and almost exclusively comprised of chickweed (*Stellaria media*). Cover crops in both fall and spring influenced the biomass of the most common weed species but did not filter species differentially. These results suggest that cover crop planting date may have a larger effect on weed community assembly than cover crop identity or functional diversity, by influencing the phenological patterns of weed germination.

DEPLOYING MICROBES AS A SEED TREATMENT FOR PROTECTION AGAINST SOIL-BORNE PLANT PATHOGENS. R. Carr*, Rodale Institute, Kutztown, PA (180)

ABSTRACT

Plant diseases, especially those caused by soil-borne seed infecting pathogens are a constant threat to both greenhouse and field production of many agricultural crops. Conventional farming operations often use fumigants and chemical seed treatments for controlling seed and seedling pathogens. However, these materials can be harmful to human health and the environment. The use of many of these materials is also strictly prohibited in organic agriculture, limiting the options for plant disease control. Organic amendments such as compost and vermicompost are often used as alternatives to synthetic materials due, in part, to their success in controlling plant pathogens. Previous studies have confirmed consistent disease suppression using solid and liquid forms of organic amendments and the working hypothesis is that microbes are closely associated with suppression. Furthermore, only a subset of the microbes present in the bulk material that colonize the seed coat are responsible for disease suppression. So, if the specific subset of microbes associated with seed colonization and suppression can be deployed as a seed treatment, can we still achieve plant protection from soil-borne plant pathogens? In addition, can this seed treatment application be developed for organic production as and effective tool for disease management? This presentation will review a project in which microbes from liquid compost extracts are freeze-dried into a powder and then applied to the seed surface using standard seed treating technologies. Treated seeds were evaluated for suppression of Pythium damping-off in cucumber under laboratory conditions. From a scientific standpoint, freeze-dried extracts present a novel approach for deploying microbes for plant disease control. For organic agricultural practitioners, seed applications of freeze-dried extracts could provide an additional tool in the plant disease management toolbox.

EFFECTS OF TILLAGE AND NUTRIENT MANAGEMENT DURING THE TRANSITION TO ORGANIC GRAIN PRODUCTION. M.R. Ryan*, M. Ball, B. Caldwell, A. DiTommaso, L. Drinkwater, C. Marschner, and C.L. Mohler, Cornell University, Ithaca, NY (181)

ABSTRACT

A long-term experiment was initiated in 2005 at the Cornell University Musgrave Research Farm and compared four organic cropping systems that differed in soil fertility inputs and intensity of weed management. Soil fertility management in the High Input system was based on soil testing and Cornell University soil fertility recommendations, whereas fewer inputs were used in the Low Input system. Standard physical and cultural weed management including inter-row cultivation and delayed planting was applied in both of these systems. Additional weed management practices were applied in the Intensive Weed Management system, which also received low soil fertility inputs. The Reduced Tillage system was designed to maximize soil health and received fewer soil tillage and cultivation practices. A three year rotation of corn, soy and spelt/red clover was grown in all systems and the experiment included two crop rotation entry points, enabling two of the three crops in the rotation to be grown every year. Crop performance and weed suppression data were collected annually in all systems and crops. Results show that spelt played an important role in reducing weed populations by limiting seed production of annual weeds. Perennial weeds increased in the Reduced Tillage system, which contributed to its poor performance relative to the three other systems. Overall our findings indicate that managing soil fertility and avoiding nutrient overabundance in organic cropping systems is important for suppressing weeds and maximizing profitability.

EVALUATION OF NATIVE NEW YORK ENTOMOPATHOGENIC NEMATODES FOR BIOCONTROL OF PLUM CURCULIO (*CONOTRACHELUS NENUPHAR*) IN APPLE ORCHARDS. T.G. Lessord*, A. Agnello, E. Shields, and K. Wickings, Cornell University, Geneva, NY (183)

ABSTRACT

Plum Curculio (*Conotrachelus nenuphar*) is a prevalent economically damaging pest in tree fruit systems. This insect is common in the United States east of the Rocky Mountains. Due to increased restrictions on chemical insecticides, efforts were made to find an effective method of biological control. Two New York-native strains of entomopathogenic nematodes (EPNs), *Steinternema carpocapsae* "NY-001" and *Steinernema feltiae* "NY-04", collected in 1990, were determined to persist in New York crop systems for several years. Intact soil columns were taken in to the lab and held at constant moisture and temperature. The virulence and persistence of these two strains was tested in the given soil conditions in four apple orchard sites. While it was found that the two EPN strains were present in the soil and caused significant mortality to sentinel Galleria mellonella larvae, there was no difference in plum curculio larval survival and subsequent adult emergence between EPN-treated vs. untreated soil, indicating that the EPNs may not have been able to locate and kill plum curculio larvae under these conditions.

WEED SEED SUPPRESSION USING DESIGNED COMPOST EXTRACTS. G.M. Zinati*, Rodale Institute, Kutztown, PA (185)

ABSTRACT

Organic farmers consider weeds their number one problem in crop production. They are constantly on the lookout for practical technologies that reduce competition of weeds with crop growth. A research study was established to explore a proof of concept whether the use of designed compost extracts varying in dilutions and chemical and biological properties, would lead to reduction in weed seed expression without impacting crop seed germination under controlled conditions. Two compost piles C1 and C2 were prepared with same feedstock with varying rates to encourage microbial diversity. Compost pile 1 had 30 percent brown, 50 percent green and 20 percent high nitrogen (N) whereas, C2 had 50 percent brown, 30 percent green and 20 percent high N. Both C1 and C2 were analyzed for chemical and microbial diversity. Compost Extract (CE) dilutions (1:2, 1:3, 1:4, and 1:5 (w/v) were prepared from each. Ten seeds each of pigweed, lambsquarter, chickweed, and cucumber were placed in petridishes, lined with Whatman filter paper #1, and received 10 ml of each dilution. Seeds were incubated for 5 days at 24 ^{o}C and compared to those in deionized water (DI) in randomized complete block (RCB) design with four replications. Under controlled conditions, seeds from each of weeds and crops were placed in pots containing autoclaved sand and received 35 ml of compost extracts at 1:3 dilution or DI water. An increase in dilution ratio from 1:2 to 1:5 resulted in significant and linear decrease in EC and NO_3 concentration. In petridish test C1 and C2 extracts increased significantly cucumber root length by 37 percent and 34 percent, respectively when compared to DI water. Pigweed seed germination was reduced by 20 percent when treated with C1 extract. Similar results were obtained for cucumber and pigweed seeds when grown in pots containing autoclaved sand under controlled conditions. Compost extracts at 1:3 dilution prepared from compost with feedstocks, as described in C1, can be considered better for pigweed suppression as they contain lower levels of P, Ca, Mg, Fe, Cu, and Zn, protozoa and nematode.

AGRONOMIC CROPS (I)

IS AN INTEGRATED WEED MANAGEMENT RENAISSANCE OR FALLACY IN OUR FUTURE? W.S. Curran*, J.M. Wallace, D. Lingenfelter, D.A. Mortensen, and M.J. VanGessel, Penn State, University Park, PA (186)

ABSTRACT

This presentation could perhaps be better titled "can herbicide resistance drive adoption of integrated weed management"? This has been an active discussion within the weed science community over the last several years with little progress and we are experiencing an herbicide resistant weed crisis. According to Heap (Nov. 27, 2015), there are 461 unique cases of herbicide resistant weeds globally with 247 species. Further, weeds have evolved resistance to 22 of the 25 known herbicide sites of action and to 157 different herbicide active ingredients. All weed management practices are selective tools and we should never recommend the excessive use or over reliance of any selective tool. Rabb in 1972 defined "integrated pest management" as the intelligent selection and use of pest control actions that will ensure favorable economic, ecological, and sociological consequences. This concept evolved because of insecticide resistance and due to other environmental concerns with pesticides. It has never been widely practiced in the science of managing weeds and perhaps less now than ever. Weed management in most of our major field crops is almost exclusively herbicide-based. No-till, Roundup Ready, cost effective herbicides, and simplicity help drive this norm. At a weed science conference a few years back, Steve Shirtliffe (University of Saskatchewan) acknowledged "how can increased complexity be sold when simplicity is craved". Bagged lettuce, Keurig coffee, and rotisserie chickens are every day examples of how simplicity and convenience are universal and part of our everyday lives. Why shouldn't weed control have the same convenience and simplicity? Unfortunately, because NATURE always wins! The United States leads the world in herbicide resistance - 153 unique cases with examples of resistance in most herbicide families. What should be the role of IPM in managing weeds in 2016 and beyond? What is our responsibility as scientists and educators in making sound weed control recommendations? This presentation will revisit some IPM options and discuss some of our research that has investigated integrating nonchemical tactics or reduced herbicide strategies for diversifying weed management.

AN INTEGRATED WEED MANAGEMENT APPROACH TO ADDRESSING THE MULTIPLE HERBICIDE-RESISTANT WEED EPIDEMIC IN THREE MAJOR U.S. FIELD CROP PRODUCTION REGIONS. S. Mirsky*, A.S. Davis, J. Norsworthy, M. Bagavathiannan, J. Bond, K. Bradley, W. Curran, W.J. Everman, D. Ervin, M.L. Flessner, G. Frisvold, A. Hager, B. Hartzler, N. Jordan, J. Lindquist, B. Schulz, L. Steckel, and M.J. VanGessel, USDA-ARS, Beltsville, MD (187)

ABSTRACT

Multiple herbicide-resistant (MHR) weeds are challenging sustainable crop production (reduced- and no-tillage production) as the herbicides that farmers have relied on for decades to control weeds are rapidly becoming less effective and the pace of herbicide discovery has greatly slowed. To manage MHR weeds successfully in the future, farmers need to employ multiple control tactics within an integrated weed management approach. One promising tactic for managing MHR weeds is Harvest-time Weed Seed Control (HWSC), in which weed seeds are removed or destroyed at the time of harvest. Previously, only hand-weeding was an option, but new mechanical ways of accomplishing HWSC (chaff carts, narrow windrows, and Harrington Seed Destructor) are starting to be developed. Preliminary field trials and simulation models show that HWSC can be particularly effective against annual weeds, which dominate most of our cropping systems and represent the primary economic impact of MHR weeds. Our team proposes to quantify the potential for integrated weed management systems to help producers regain control of MHR weeds in the north central, south central and mid-Atlantic U.S grain production regions while preserving over a half century of progress toward reduced tillage. These systems will include the following control tactics: herbicides (chemical); cover crops (physical, biological, and chemical); and HWSC (physical). We will use a core-satellite design: at core sites, project directors will lead comprehensive studies of the impact of integrated weed management on population and resistance dynamics of MHR weeds; at satellite locations, regional collaborators will collect supporting weed ecology and biology data to help extrapolate core site results at regional and multi-regional levels, and conduct outreach to stakeholders. Economic and social information critical to the adoption of integrated weed management systems will be developed. Outreach efforts will include providing web-based content, conducting and participating in workshops and farmer field days, and "peer-to-peer" learning through community-based programs to promote hard (weed control tactics) and soft (scouting) technology adoption. As we learn more about the efficacy of integrated weed management for addressing MHR problems, we will deliver a coherent, compelling outreach message to help producers make management decisions to meet this challenge.

EVALUATION OF DICAMBA-BASED SYSTEMS FOR CONTROL OF GLYPHOSATE-RESISTANT PALMER AMARANTH. C.W. Cahoon*, A.C. York, S. Culpepper, D.L. Jordan, W.J. Everman, and K.M. Jennings, Virginia Tech, Painter, VA (188)

ABSTRACT

Glyphosate-resistant (GR) Palmer amaranth (AMAPA) remains troublesome throughout the southern United States. To aid in the control of this weed, Monsanto has developed cotton and soybean cultivars with tolerance to glufosinate, glyphosate, and dicamba and glyphosate and dicamba, respectively. Dicamba offers wide spectrum broadleaf control and will be an additional postemergence (POST) over-the-top and preemergence (PRE) option for weed control in both crops. An experiment was conducted at North Carolina during 2013 and 2014 and Georgia during 2013 to evaluate AMAPA control with herbicide systems including dicamba. Soil at each field sites was loamy sand with 0.46 to 1.9 percent organic matter and greater than 100 AMAPA/m^2. The experiment consisted of a factorial treatment arrangement of two base herbicide systems and seven timings of dicamba. All plots received acetochlor (1260 g ai/ha) prior to emergence of Palmer amaranth (PRE). The two base herbicide systems were glyphosate potassium salt (1260 g ae/ha) and glufosinate-ammonium (654 g ai/ha). These herbicides were applied to Palmer amaranth 10 cm or less in height 18 to 23 days after PRE (POST 1) and 18 to 22 days later to 10 to 40 cm tall Palmer amaranth (POST 2). Size of Palmer amaranth at POST 2 depended upon the previous herbicide applications. Timing of dicamba applications included no dicamba, PRE, POST 1, POST 2, PRE and POST 1, PRE and POST 2, and POST 1 and POST 2. Data for AMAPA control were subjected to ANOVA using the PROC GLIMMIX procedure in SAS and means separated using Fisher's Protected LSD at p =0.05. Acetochlor alone PRE controlled AMAPA 59 to 83 percent prior to POST 1. At all locations, AMAPA control by acetochlor plus dicamba was greater than acetochlor alone (78 to 99 percent). Dicamba benefited both the glyphosate and glufosinate system. However, because of the significant amount of GR AMAPA at each location, the increase in control was more dramatic in the glyphosate system. Late in the season, glufosinate and glyphosate applied alone POST 1 and POST 2 controlled AMAPA 76 and 32 percent, respectively. Adding dicamba PRE or POST 1 in the glufosinate system did not improve AMAPA control compared to glufosinate alone. The addition of dicamba at POST 2 to glufosinate increased AMAPA control 15 percent. Likewise, two applications of dicamba in the glufosinate system improved AMAPA control 19 to 23 percent. In the glyphosate based system, dicamba added PRE, POST 1, or POST 2 increased AMAPA control 19, 47, and 50 percent, respectively. Dicamba applied PRE and either POST 1 or POST 2 also improve AMAPA control. However, in the glyphosate system, the greatest increase in control of AMAPA was observed following dicamba applied twice POST. When AMAPA were large (20 to 25 cm), dicamba applied POST improved AMAPA control by glufosinate and glyphosate alone 27 and 70 percent, respectively. In general, including dicamba with glufosinate or glyphosate increased AMAPA control. However, increases in AMAPA control were much more noticeable when dicamba was added to glyphosate. Furthermore, glufosinate alone controlled AMAPA greater than glyphosate alone. This was to be expected with the high level of GR AMAPA. However, when dicamba was included with each of these herbicides, differences in weed control by glufosinate and glyphosate were only minor. Therefore, where GR weeds are present, glyphosate plus dicamba may prove to be a useful weed control option. Furthermore, dicamba combined with glufosinate greatly increased AMAPA control when applied to AMAPA greater than 10 cm in height. Growers will find utility in adding dicamba to glufosinate when weeds are larger than desired for glufosinate alone to control. Additionally, tank mixes of dicamba plus glufosinate include two modes of action, reducing the chances of weed resistance developing to either herbicide.

NUTRIENT MANAGEMENT IN MID-ATLANTIC ORGANIC GRAIN CROPPING SYSTEMS. V. Ackroyd*, S. Mirsky, M.A. Cavigelli, and J.T. Spargo, USDA-ARS, Beltsville, MD (189)

ABSTRACT

Farmers typically use animal manure or compost to maintain soil fertility. While these materials have N:P ratios of approximately 2:1 and 1:2, respectively, most crops have ratios between 7:1 and 10:1. Manure applications sufficient to meet crop N needs thus typically provide excessive amounts of P, which, with repeated applications, can result in increased risk of P losses to surface runoff. In watersheds such as that of the Chesapeake Bay, the P load of surface runoff is of such environmental concern that manure application is regulated at both the state and federal levels. In addition, manure is a costly nutrient source with limited availability. Farmers have turned to supplementary nutrient sources, such as legume cover crops, to supply the needed N without adding excessive P to the system. Furthermore, it appears that fields with a long history of manure application have extensive soil reserves of both P and N. The purpose of this research was to determine whether manure applications could be decreased when used in conjunction with legume cover crops, without impacting crop yield. In fields with a history of manure applications, rates can be reduced without negatively impacting corn yield.

POSTEMERGENCE WEED CONTROL IN CORN USING VARIOUS RATES OF ATRAZINE AND TOPRAMEZONE. K.M. Vollmer* and T. Hines, University of Delaware, Georgetown, DE (191)

ABSTRACT

The p-hydroxyphenyl pyruvate dioxygenase (HPPD)-inhibiting herbicides have been shown to be effective in controlling many grass and broadleaf weed species. The effectiveness of this mode-of-action is has been shown to improve with the addition of atrazine to the tank mix. Label recommendations suggest applying topramezone at 18.4 g ha^{-1} with 560 ha^{-1} atrazine as a postemergence treatment for control of several broadleaf and grass species. The objective of this study was to determine if the recommended rates of the topramezone/atrazine tank mix could be reduced by half while maintaining the same weed control efficacy. In 2013 and 2014, field trials were established in conventional–tillage corn, and in 2015 in a no-tillage corn at the Virginia Tech Eastern Shore Agricultural Research and Extension Center in Painter, VA. The study was a 2 by 3 factorial with herbicide and rate as main factors arranged in a randomized complete block design with 3 replications. Factors were herbicides, topramezone or atrazine, and rate, 0, 0.5, and 1X. In 2015 a preplant burndown treatment of atrazine plus glyphosate applied at 1,121 and 1,065 g ai $ha^{-1}$was applied two weeks before planting. Treatments were applied postemergence, when weeds reached 10 to15 cm in height. Treatments consisted of topramezone applied at 0, 9.2 (0.5X), and 18.4 (1X) g ha^{-1} and atrazine at 0, 280 (0.5X), and 560 (1X) g ha^{-1}. Treatments included methylated soybean oil and urea ammonium nitrate at 1 percent and 1.25 percent v/v respectively. Plots were visually evaluated for percent control 7 days after the postemergence treatment on a scale of 0 to 100 with 0 being no control and 100 being complete weed control.

Weeds in the conventional tillage plots consisted of common lambsquarters, common ragweed, ivyleaf morningglory, and smooth pigweed. Bermudagrass, giant foxtail, ivyleaf morningglory, and large crabgrass were prevalent in the no-tillage plots. Therefore, data were separated by tillage system. In the conventional tillage trials there were no significant year by treatment interactions for smooth pigweed control or common lambsquarters control so data was pooled over years. All topramezone/atrazine combinations controlled common lambsquarters, common ragweed, ivyleaf morningglory, and smooth pigweed at least 94 percent both years. In 2014, common ragweed control with topramezone at the 1X alone was similar to topramezone/atrazine combinations. The topramezone/atrazine combination provided better control of ivyleaf morningglory than topramezone alone. In 2015 in the no-tillage trial, both rates of topramezone and all topramezone/atrazine combinations controlled bermudagrass, large crabgrass, giant foxtail, and ivyleaf morningglory at least 92 percent when rated 7 days after treatment. Control of these species declined with later rating dates, due to additional emergence of these species. This was reflected in lower average corn yields (1,643 kg ha^{-1}) compared to the conventional tillage studies in 2013 (9,451 kg ha-1) and 2014 (5,240 kg ha^{-1}). The recommended rates for the topramezone plus atrazine tank mix can be reduced by half without a significant reduction in the postemergence control of certain weed species. However, these treatments did not provide sufficient residual grass control in no-tillage systems, indicating that an additional herbicide is warranted.

TURF AND ORNAMENTALS

IPM IN YOUR POCKET: THE GREENHOUSE SCOUT MOBILE APP. E.M. Lamb* and B.C. Eshenaur, New York State Integrated Pest Management, Ithaca, NY (192)

ABSTRACT

Greenhouse growers ask us most often for information on pest identification and biocontrol methods, and we most often wish they would scout consistently and keep good records. To ease all those activities, we created a mobile application called Greenhouse Scout. Available on Apple and Android devices, it offers quick access to information on most greenhouse insect and mite pests, including photographs, biology and monitoring methods. For beneficial insects, mites and nematodes, there is information on biology, checking for viability, application methods, and interactions with pesticides. The interactive portion of the app allows growers to record scouting information and applications of beneficials and pesticides. This data can be viewed as it was entered or in graphical form, to see changes in the pest population. Information is archived or can be printed in a variety of forms. Growers do not need to be connected to the internet to input data, as the app will sync with the website as soon as a connection is detected.

POLITICS, PLANT HEALTH & PUBLIC PERCEPTION. J.L. Fetzer*, M-NCPPC, Gaithersburg, MD (7)

ABSTRACT

Lessons learned from Bill 52-14 Cosmetic Pesticide Use Restrictions in Montgomery County Maryland. Politicians are concerned with the physical and economic well-being of their constituents but may not understand or value the critical relationship between healthy ecosystems, plants and human health. Scientists and plant care practitioners must engage in conversations with policy makers, but often speak a different language. After many meetings with policy-makers; testifying at the county and state level; and conversations with colleagues in large county parks systems... What would we do the same? What lessons were learned to suggest different choices of words and messaging?

We learned that it was essential to communicate with colleagues across the US & Canada under similar prohibitions—getting details, especially fiscal impacts, success, and failures created a powerful statement in conversations with policy makers. Early on engage upper management in your agency and get input from managers and staff who would be directly affected by proposed legislation. Build relationships with local agencies, business operators, and others because working as a team helps better inform the issues and potential impacts. Connecting with reputable news reporters and social media can be a good tool for communicating opposing views on heated issues.

Our science can only be important if we share it! Communication is most effective when using broadly relatable language and examples that can establish a direct personal connection. Images should convey data in a simple, step-by-step, understandable way. Our focus is on explaining science so policy makers make informed decisions—it is NOT to persuade. Consider coaches especially for those who testify; it is helpful to get tips on messaging, how to handle difficult issues and tools for keeping cool under pressure.

Bill 52-14 did pass with amendments and changes that were being added up to the final minutes in the voting session. It was important to work with staff on rewriting bill language and definitions; it was also essential to maintain on-going contact with individual policy makers. The political procedures can be confusing and the protocols can vary at the local vs state levels; a lobbyist can help inform you of what to expect in various political arenas. We CAN make a difference in public policy if we step forward to effectively communicate our science.

MAKING AECIAL-TELIAL HOST CONNECTIONS: NEW INSIGHT INTO HOST ALTERNATION OF THE PEAR TRELLIS AND JAPANESE APPLE RUST FUNGI IN THE NORTHEAST. S.C. Kenaley and M.L. Daughtrey*, Cornell University, Riverhead, NY (8)

ABSTRACT

Pear trellis rust caused by Gymnosporangium sabinae was first reported from the northeastern United States in 2012, affecting only the rosaceous (aecial) hosts Pyrus communis and several cultivars of P. calleryana in central and southeastern New York. To date, the infection of junipers (Juniperus, Section Sabina; telial hosts) in the U.S. has been noted only in California. Similarly, the Japanese apple rust fungus (G. yamadae) was first noted in 2004 on crabapple (Malus toringo) in the eastern U.S., whereas the telial stage has been reported in the U.S. only once, in 2009, on J. chinensis in Delaware. Scouting efforts in spring 2015 in Suffolk County, NY detected telia on nursery- and landscape-grown Juniperus, including Juniperus chinensis 'Pfitzeriana' and J. chinensis 'Robusta Green'. Leaves of crabapples (Malus spp.) and apples (Malus x domestica) with conspicuous aecia were also collected in the region. Molecular (LSU D1-D2 domain) and morphological analyses of the aecial and telial stages revealed two rusts – G. sabinae and G. yamadae – producing telia on stems of J. chinensis 'Robusta Green', while telia on Juniperus chinensis 'Pfitzeriana' were determined to be G. yamadae. The latter rust fungus was also confirmed on crabapple and edible apple within the scouted area. Thus, these and other junipers in the Section Sabina likely have played a critical role in the introduction and establishment of pear trellis and Japanese apple rust fungi within the region.

HORSEWEED CONTROL IN FIELD NURSERY CROPS. J. Altland*, USDA-ARS, Wooster, OH (196)

ABSTRACT

Horseweed (*Conyza canadensis*) is one of the most problematic weeds across all crops in agriculture. Its economic impact worsened when it developed resistance to glyphosate-containing herbicides. Nursery growers in the Midwest have recently reported difficulty in controlling horseweed in field-nursery crops with glyphosate applications. Considering this, the only practical method for controlling horseweed in nursery crops is through an effective preemergence program. However, very few of the preemergence herbicides registered for use in nursery crops have been evaluated in the scientific literature for horseweed control. The objective of this research was to evaluate preemergence herbicides for horseweed control in a field soil.

Herbicides were spray-applied to a Canfield silt loam soil on October 1, 2014, in plots 1.2 m wide and 3 m long. Herbicides were applied with a CO_2 backpack sprayer equipped with a two-nozzle boom and calibrated to deliver 374 L/ha. The following herbicides were applied: isoxaben at 1.12 kg/ha ai, flumioxazin at 0.42 kg/ha ai, oxyfluorfen at 1.68 kg/ha ai, indaziflam at 0.084 kg/ha ai, simazine at 4.48 kg/ha ai, pendimethalin at 5.1 kg/ha ai, and dimethenamid at 1.68 kg/ha ai. A non-treated control was also included. The field in which the experiment was conducted has a history of heavy horseweed pressure. Nonetheless, approximately 2000 horseweed seed collected from local populations were applied to each plot in a seed and sand mixture. There were five replications per treatment arranged in a randomized complete block design. Weed counts were conducted the following spring to assess control of horseweed and other weed species. Weed counts were conducted by randomly tossing a 30-cm diameter copper ring (0.07 m^2 area) into each plot three times, and counting the number of each species within the ring.

In addition to horseweed, the most prevalent weed species observed in plots included annual bluegrass (*Poa annua*), lambsquarter (*Chenopodium album*), and Pennsylvania smartweed (*Polygonum pensylvanicum*). Isoxaben, indaziflam, and flumioxazin provided the most effect horseweed control, each with less than 1 weed per 0.07 m^2 compared to more than 11 horseweed per 0.07 m^2 subplot in non-treated controls. Oxyfluorfen, pendimethalin, and simazine reduced horseweed numbers compared to non-treated controls, although not as effectively as the three aforementioned herbicides. Dimethenamid application resulted in similar horseweed numbers to non-treated controls. While isoxaben provided effective control of horseweed and other broadleaf weeds, it provided no control of grass weeds. Indaziflam and flumioxazin provided nearly 100 percent control of all weed types. Pendimethalin provided effective grass control, but moderate to poor control of horseweed and other broadleaf weeds.

Results from this experiment suggest that flumioxazin and indaziflam would provide excellent broad-spectrum weed control, including horseweed, throughout spring and early summer when applied in a fall application. A tank mix of isoxaben, which provided good horseweed and broadleaf weed control, and pendimethalin, which provided excellent grass control, might in combination provide effective broad spectrum weed control in sites where phytotoxicity is a concern with flumioxazin or indaziflam.

DORMANT APPLICATIONS OF MARENGO SC AND G COMPARED TO AN INDUSTRY STANDARD IN A NURSERY OVERWINTERING POLYHOUSE. H.M. Mathers*, Mathers Environmental Science Services, LLC, Gahanna, OH (197)

ABSTRACT

The practice of dormant pre-emergence herbicide applications in nursery overwintering containers is becoming more common. There are no registered preemergence herbicides for this use. Marengo SC (Indaziflam 7.4 percent by wt.) (OHP, Inc., Mainland, PA, 19451) was registered for under greenhouse benches in 2014. The sponsor of this trial, OHP, was interested in the phytotoxicity, efficacy and duration of efficacy of Marengo SC at 7.5, 15 and 30 oz./ac rates versus the Marengo G (0.0224 percent a.i. indaziflam) 200 lb./ac rate, applied OTT, to dormant containerized plants, pre-winter close up. The Marengo formulations were compared to the industry standard for these dormant applications, SureGuard WDG (*Flumioxazin 51 percent) 6 oz. /ac,* (Nufarm Americas Inc., Alsip, IL, 60803) and an untreated control. Seven species with varying levels of phytotoxicity to indaziflam and flumioxazin, in active growth, were chosen for dormant evaluations, *Spirea japonica* 'Little princess', *Stachys minor*, *Berberis thunbergii 'Concord', Juncus effuses, Rhododendron* 'Dandy Man Pink, *Hydrangea macrophylla* 'Endless Summer' and *Phlox paniculata.* The host site was Smith's Gardens, Inc., Delaware, OH, 43015. For each species and treatment there were eight single plant replicates arranged in a completely randomized design (CRD) within species and 336 total trial plants. *Spirea*, *Berberis*, and *Hydrangea* all experienced phytotoxicity above commercial acceptable $\leq$ 3 (0-10 rating, with 0 being no injury) and should not be used even dormant with Marengo SC, G or SureGuard. Marengo G 200 lb. /ac was found safe on dormant *Stachys, Juncus* and *Phlox*. This finding is significant to herbaceous perennial nursery container grower as *Stachys* and *Phlox* are two of the most sensitive herbaceous species to herbicide injury, and are currently kept weed free without chemical controls. The Marengo SC 7.5 oz./ac rate was also safe on dormant *Juncus*, *Phlox* and *Rhododendron*; however, confounding cold injury suggests the *Rhododendron* and Marengo SC 7.5 oz./ac should be evaluated further in future studies. All treatments provided significantly superior efficacy than the control, pooled over species and evaluations. All treatments provided commercially acceptable efficacy $\geq$ 7 (0-10 rating, with 10 being perfect control), except the control. Marengo SC 7.5, Marengo G and SureGuard were not statistically different in there level of efficacy but again were different in their phytotoxicity. Pooled over evaluations and species and comparing phytotoxicity and efficacy, Marengo G and Marengo SC 7.5 oz./ac are the two best treatments for dormant applications.

FUNGICIDE SENSITIVITY IN BOXWOOD BLIGHT PATHOGEN *CALONECTRIA PSEUDONAVICULATA*. K.A. Maurer* and J.A. LaMondia, The Connecticut Agricultural Experiment Station CAES, Windsor, CT (198)

ABSTRACT

Boxwood blight caused by *Calonectria pseudonaviculata* is a devastating disease of boxwood resulting in defoliation, dieback, and high economic losses. Fungicide efficacy against the pathogen was proven in previous studies. Three isolates: 11-9-4a - a sensitive wild type; FC1 - an isolate of 11-9-4a selected for ability to grow on up to 250 µg a.i./ml pyraclostrobin; and a recent isolate CTWH1 from a Connecticut landscape were evaluated *in vitro* to analyze fungicide sensitivity for four strobilurins (pyraclostrobin, azoxystrobin, trifloxystrobin, kresoxim-methyl) and five demethylation inhibitor (DMI) fungicides (propiconazole, tebuconazole, triflumizole, myclobutanil, tetraconazole). Radial growth was used to assay for fungicide sensitivity at concentrations of 0, 3.16, 10, 31.6, 100, 316, and 1000 µg a.i./ml. DMIs strongly inhibited mycelial growth for all isolates, complete inhibition was achieved with propiconazole, tebuconazole, and triflumizole at concentrations of at least 1 µg a.i./ml and with myclobutanil and tetraconazole at 10 µg a.i./ml. The 85 percent effective dose (EC_{85}) values show that pyraclostrobin and kresoxim-methyl had the highest effect on the pathogen. Fc1 and CTWH1 exhibited reduced sensitivity to strobilurin, but not DMI fungicides compared to 11-9-4a. This study shows that reduced strobilurin fungicide sensitivity in *C. pseudonaviculata* can be selected for *in vitro* and can also occur naturally in pathogenic isolates in the landscape.

CURATIVE FUNGICIDAL ACTIVITY AGAINST *CALONECTRIA PSEUDONAVICULATA*, CAUSAL AGENT OF BOXWOOD BLIGHT.

J.A. LaMondia*, The Connecticut Agricultural Experiment Station, Windsor, CT (199)

ABSTRACT

Boxwood blight, caused by *Calonectria pseudonaviculata (Cps)*, is a recently emerged disease of *Buxus* and *Pachysandra*. Selected efficacious fungicides were evaluated for post-infection activity. *In vitro*, *Cps* exposed to $\geq$ 5 ppm DMI fungicides in agar did not resume growth when transferred to non-amended media, suggesting the potential for curative activity. A detached leaf assay was used to evaluate 0, 3, 30 or 300 ppm propiconazole, tebuconazole or kresoxim-methyl on disease when applied 24 h before or 24, 48 or 72 h after inoculation (curatively) with 100 conidia per leaf. All fungicides were effective from 3 to 300 ppm when applied preventatively. Propiconazole reduced disease incidence at 30 or 300 ppm at 24 h applied curatively and prevented sporulation at 30 or 300 ppm up to 96 h. Tebuconazole and kresoxim-methyl did not reduce disease when applied curatively; tebuconazole but not kresoxim-methyl prevented sporulation at 30 or 300 ppm up to 96 h. Propiconazole applied curatively at 30 or 300 ppm reduced diseased leaf area; 300 ppm resulted in 2, 18 or 51 percent of leaf area diseased at 24, 48 or 96 h compared to 90 percent for untreated leaves. Green Velvet boxwood plants were treated with propiconazole, thiophanate-methyl, pyraclostrobin, kresoxim-methyl or water alone 48 h after inoculation with *Cps*. Disease development was reduced after curative application of propiconazole only. The number of *Cps* re-isolations from symptomatic tissue was also reduced for propiconazole.

THE ROLE OF HERBICIDE APPLICATION TIMING IN EFFICACY ON DORMANT NURSERY FIELDS THE FOLLOWING SPRING. H.M. Mathers*, Mathers Environmental Science Services, LLC, Gahanna, OH (200)

ABSTRACT

Phytotoxicity and efficacy were evaluated on three fields per each of three nursery locations in Ohio with over the top (OTT) applications on dormant growth. The locations included Studebaker Nurseries, Inc., New Carlisle, OH, 45344; *Wm A Natorp Co., Mason, OH, 45040 and Herman Losely and Son, Inc., Perry, OH, 44081.* The species at Studebaker Nurseries included boxwood (*Buxus* 'Green Velvet' - 1yr and 5yr fields) and yew (*Taxus Xmedia* **'Runyan'** – 3yr) with one application made April 1, 2015. *The* species at Natorp Co., included boxwood (*Buxus* 'Green Gem'- 3yr), (*Buxus* 'Green Velvet' - 1yr) and yew (*Taxus X media 'Densiformis'* - 1 yr.) with two applications made on similar adjacent beds (but not twice on the same beds), December 11, 2014 and March 12, 2015. The species at Losely's included yew (*Taxus* X*media* 'Tauntonii' – 3yr and *Taxus cuspidata 'Green Wave'* - 1yr) and boxwood (*Buxus sempervirens* 'Green Mountain' – 3yr) with one application made April 8, 2015. Three benefits of dormant applications are noted: 1) Utilization of nursery staff in winter which is traditionally a "down-time" labor-wise versus spring; 2) Insurance that applications will be completed before germination, versus waiting for spring when conflicting operations take precedence; and, 3) Optimized control of weeds that are active in cold weather including wild garlic, nutsedge, Canada thistle and other problematic perennial and biennial weeds. Dormant applications also seem to be key in providing the "power" and "duration" of efficacy necessary to clean up nursery fields recovering after "downturned" economic years in the industry, without increased phytotoxicity. Results indicate Casoron, Lontrel, V10233 and SureGuard were all performing better with December applications at Natorp's. This is an interesting new finding that could change how these products are advocated for nursery field and even landscape use in the future. Marengo SC was the only treatment that was better in the March applications. Losely, *Taxus* X*media* 'Tauntonii' – 3yr, was the only species at the three nurseries to exhibit any long term phytotoxicity, weeks after treatment (WAT), with three treatments: Certainty, Echelon and Lontrel. Only Certainty provided non-acceptable phytotoxicity at 14 WAT. At Studebaker, where the predominant weed was Canada thistle, Lontrel was by far the most efficacious treatment providing commercially significant weed control to 12 WAT and 10 WAT, depending on crop. Lontrel was followed by Casoron 4G providing commercially significant weed control to 10 WAT and 7 WAT, again dependent on crop, at Studebaker's. At Natorp's, Casoron CS performed the best in the December 11, 2014 applications providing commercial control 28 and 26 WAT, depending on the crop. Casoron CS was followed by Marengo SC with commercial control at 26 and 21 WAT, depending on crop. With the March 12, 2015 applications at Natorp's, Marengo SC was the most efficacious treatment providing 15 and 13 WAT commercial control depending on crop and followed by V10233 with commercial weed control 13 WAT in two of three crops. The Gallery SC + Dimension 2EW treatments performed best at Natorp's in the December 11, 2014 applications providing commercial control 16 WAT.

USE OF SYSTEMIC TRUNK INJECTION FOR CONTROL OF BLACK OAK GALL WASP ON CAPE COD, MA. D.M. Grosman*, M. Davis, and J. Elkinton, Arborjet Inc, Woburn, MA (201)

ABSTRACT

The black oak gall wasp (BOGW), *Zapatella davisae,* has recently caused dieback and mortality of black oaks in Cape Cod and Martha's Vineyard, Massachusetts and Rhode Island. There is no known sustainable control option for BOGW other than maintaining tree vigor. Recently, an injected systemic insecticide, containing emamectin benzoate (TREE-age™; Arborjet Inc., Woburn, MA) has shown some promise for improving the health of trees being attacked by stem gall wasps. Two separate trials were established in fall 2013 and spring 2014 to evaluate efficacy of recommended rates of emamectin benzoate (TREE-age™) or imidacloprid (IMA-jet) for improving the health of individual black oak trees infested with BOGW. As of 13 - 21 months post-treatment, both insecticide treatments, have significantly reduced adult emergence and improved tree health compared to untreated checks. These studies are ongoing and this poster presents preliminary data only.

EFFECTS OF FUNGICIDES ON SPORE GERMINATION AND COLONY DEVELOPMENT OF *ERYSIPHE PULCHRA*, THE CAUSAL AGENT OF POWDERY MILDEW OF FLOWERING DOGWOOD. Y. Li*, M. Windham, R. Trigiano, D. Fare, and W. Copes, The Connecticut Agricultural Experiment Station, New Haven, CT (202)

ABSTRACT

Powdery mildew caused by *Erysiphe pulchra* is a destructive disease in flowering dogwoods. Application of fungicides is an important component in the integrated pest management program of powdery mildew diseases. Effects of five fungicides (propiconazole, thiophanate-methyl, iprodione, azoxystrobin, and chlorothalonil) on conidial germination, infection structure formation, and colony development of *E. pulchra* were studied using a leaf disk bioassay in a laboratory. Fungicides were applied at 24 h before or after inoculation, which were defined as pre-inoculation and post-inoculation treatments, respectively. Compared to the water control, all tested fungicides significantly inhibited spore germination and decreased percent germinated conidia with branched hyphae in the pre-inoculation treatment. However, in the post-inoculation treatment, percentages of spore germination and germinated conidia with branched hyphae for all fungicides were not significantly different from water control with the exception of the significant reduction of percent germinated conidia with branched hyphae for propiconazole and thiophanate-methyl. Inhibition levels of spore germination and branch hyphae development for chlorothalonil were significantly lower than the other four fungicides in the pre-inoculation treatment. All fungicides significantly decreased disease severity both in pre- and post-inoculation treatments.

SUSCEPTIBILITY OF *IMPATIENS* SPECIES TO DOWNY MILDEW CAUSED BY *PLASMOPARA OBDUCENS*. J.E. Keach, M.L. Daughtrey*, M.P. Bridgen, and C. Salgado, Cornell University, Riverhead, NY (203)

ABSTRACT

Records of downy mildew caused by *Plasmopara obducens* on *Impatiens* in North America go back to 1881, when it was collected on the native jewelweed, *I. fulva* (syn. *I. capensis*), in Massachusetts. The disease became relevant to the U.S. bedding plant industry beginning in 2004, followed by widespread landscape losses of impatiens in 2011-12 that have continued to the present. The threat of the disease has significantly reduced use of *I. walleriana* as a bedding plant. A study in Riverhead, NY in summer 2012 exposed 18 species of *Impatiens* to downy mildew inoculum from infected *I. walleriana* in June. Potted plants were given overhead irrigation twice daily in a shaded hoop house. Several weeks to months later, sporulation of *P. obducens* was observed on known hosts *I. walleriana, I. capensis* and *I. balsamina*, and also on *I. auricoma*, *I. arguta*, *I. flanaganae* and *I. hochstetteri*. A study in summer 2015 used a similar set of conditions to challenge 23 *Impatiens* species with downy mildew. Observations of symptoms and sporulation were supported by sequencing the rDNA ITS region, which showed 100 percent nucleotide identity with *P. obducens* on *I. walleriana* for a number of new hosts. Infected *I. briartii*, *I. cinnabarina*, *I. grandis*, *I. irvingii*, *I. laurentii*, *I. repens* and *I. sodenii* var. *uguensis* showed scattered leaf lesions, and some species showed stem discoloration, but none of these new hosts were defoliated in the manner of the highly susceptible *I. walleriana*.

EFFECT OF NEW BLOCK CO-POLYMER BLENDED WETTING AGENTS ON WETTABILITY OF PEAT-BASED SUBSTRATE AND ON RELATIVE PHYTOTOXICITY OF IMPATIENS AND PANSY. By M.W. Olszewski, J. Pils, and B. Snyder, Temple University, Ambler, PA. (204)

ABSTRACT

New wetting agent formulations consisting of proprietary blends containing sulfonic acid ester ethylene oxide/propylene oxide block copolymer blend (ACA3204-R and ACA3204-P); sulfonic acid ester ethylene oxide/propylene oxide block copolymer blend with polyethylene glycol addition (ACA3204-2A), and AG2000M (Aquatrols Corporation, Paulsboro, NJ; 99% ethoxylated alkylphenol; 1% water) were tested for wettability of peat-based substrates and for phytotoxicity. Comparative wettability of peat-based substrate was evaluated following three wetting cycles using parameters of leaching fraction (percentage of water leached following an irrigation event) and wettability ratings (1 = poor wetting capability; 5 = excellent wetting capability). Substrate incorporation rates used in this study were 116 $mL{\cdot}yd^{-3}$ (a low recommended rate), 232 $mLyd^{-3}$ (a moderate recommended rate), and 464 $mL{\cdot}yd^{-3}$ (a supra-optimal rate). After a third wetting cycle, those substrates incorporated with ACA3204-R, ACA3204-P, or AG2000M at 232 $mLyd^{-3}$ had higher wettability ratings than Pro-Mix HP (Premier Horticulture Inc., Quakertown, PA) control. Following wetting agent incorporations into substrates, there were no impatiens (*Impatiens walleriana*) shoot dry weight (SDW) or visual root health rating (VRHR) differences between Pro-Mix HP and other treatments. For pansy (*Viola* × *wittrockiana*), ACA3204-R, ACA3204-P, or ACA3204-2A incorporations into peat-based substrate resulted in decreased VRHR at 464 $mL \cdot yd^{-3}$ compared to the control. There were no SDW or VRHR differences between Pro-Mix HP and AG2000M-treated substrate at any rate for either pansy or impatiens. Following soaking blotters in 150 to 1200 ppm (vol:vol) solutions of ACA3204-R, ACA3204-P, or ACA3204-2A, it was determined that solutions had no effect on final germination percentage of pansy or impatiens compared to distilled water control. In general, higher rates of wetting agent increased mean days to 50% germination (D_X; an inverse measure of germination rate) compared to distilled water control. In conclusion, new wetting agent formulations were comparatively non-phytotoxic at moderate rates for substrate incorporation and at higher rates for seed germination effects; ACA3204-R had similar wettability rating to that of AG2000M following the third wetting cycle.

POTENTIAL OPTIONS FOR SELECTIVE CONTROL OF LIVERWORT (*MARCHANTIA POLYMORPHA*) IN NURSERY CROPS. J. Derr* and A. Rana, Virginia Tech, Virginia Beach, VA (205)

ABSTRACT

Liverwort (*Marchantia polymorpha*) continues to be a problem weed in propagation of nursery stock as well as in container production. There currently are no selective controls for removing emerged liverwort from broadleaf nursery crops in outdoor production, and no selective controls for use in greenhouses and other enclosed structures. Experiments were conducted to evaluate sodium bicarbonate (baking soda) and sodium carbonate (washing soda) for control of actively-growing liverwort. A study was conducted to evaluate rates of sodium bicarbonate applied dry at 200, 400, and 800 lb/A. These three rates controlled liverwort 55, 96, and 100 percent, respectively, at 3 DAT with essentially the same control observed at 11 DAT. No significant injury to hosta (*Hosta* x 'Fire and Ice'), coral bells (*Heuchera villosa 'Miracle')* or salvia (*Salvia nemorosa* 'Maynight)' was seen in that trial. Another trial compared dry and sprayed applications of sodium bicarbonate and sodium carbonate, as well as high water volume applications to simulate applications in irrigation water. When applied dry, sodium bicarbonate and sodium carbonate controlled liverwort 74 and 81 percent, respectively, 2 WAT when applied at 200 lb/A. Sodium bicarbonate sprayed at 180 and 360 lb/A controlled liverwort 25 and 64 percent, respectively, at 2 WAT while sodium carbonate sprayed at 200 and 400 lb/A controlled liverwort 79 and 68 percent, respectively. Applying either chemical at 200 lb/A in 1,740 gal/A provided 21 percent or less control at 2 WAT. No significant injury to azalea *(Rhododendron x obtusum)*, *Nandina domestica*, sweet viburnum (*Viburnum odoratissium)*, or 'Blue Hawaii' hosta was seen in that trial. Sodium bicarbonate and sodium carbonate offer the potential for selective control of emerged liverwort from either dry or sprayed applications, although additional testing is needed to ensure nursery crop tolerance.

HERBICIDE TOLERANCE OF SEDUM SPECIES GROWN FOR GREEN ROOF INSTALLATION. A. Senesac* and N. Catlin, Cornell Cooperative Extension of Suffolk Co., Riverhead, NY (206)

ABSTRACT

Production of sedum as component species in 'green roof' tiles is a new segment of the green industry in the Northeast. Typically growers will start flats with multi-species unrooted sedum stems in a covered greenhouse. Once rooted, the flats are moved outdoors for establishment and maturation. This period outdoors is one when the flats are exposed to infestation of wind- and bird-borne weed seeds. The goal of this study was to determine if there are commercially available herbicides which can provide acceptable weed control with minimal crop damage during this period. The study was conducted at a commercial grower's site. The treated flats contained the following mixture of species and cultivars: Sedum spurium 'voodoo', S. spurium 'fuldaglow', S. spurium 'tricolor', S. spurium 'summer glory', S. rupestre 'angelina', S. sexangulare, S. acre, S. hybridum 'immergrunchen', S. kamtschaticum 'variegatum', S. pachyclados and S. sieboldii. Treatments consisted of two rates each of three granular herbicides: pendimethalin (Pendulum 2G), oxadiazon & pendimethalin (Jewel 2+1.25G) and oxyfluorfen & prodiamine (Biathlon 2+0.75G). The postemergence graminicide clethodim (EnvoyPlus) was also evaluated. The treatments were visually evaluated for percent ground cover and plant vigor and for control of overseeded mouseear chickweed (Cerastium fontanum ssp. vulgare) and annual bluegrass (Poa annua). The results suggest that, although none of the herbicides caused rapid or noticeable injury, it did appear that pendimethalin had the least negative effect on percent groundcover. Both weed species were well controlled by pendimethalin as well. It appeared that there may be a differential response of the sedum species to the various treatments. Follow up studies are being conducted to further examine this possible effect.

A NEW FUNGICIDE FOR ORNAMENTAL PLANT PROTECTION. R.J. Keese*, J.H. O'Barr, and K.E. Kalmowitz, BASF Corp., Research Triangle Park, NC (207)

ABSTRACT

BAS703F, containing the active ingredients fluxapyroxad and pyraclostrobin, will be a new tool for ornamental production. This product is a combination of two modes of action: a succinate-dehydrogenase inhibitor (SDHI, FRAC Group7) and quinone outside inhibitor (QoI, FRAC Group 11). Fluxapyroxad will be a new active ingredient in this market. This is a suspension concentrate (SC) formulation containing 2.09 pounds of fluxapyroxad and 2.09 pounds of pyraclostrobin per gallon.

Proposed use sites include greenhouses, lathhouses and shadehouses, outdoor nurseries, forest and conifer nurseries, retail nurseries and ornamentals in landscapes. The range of rates tested for ornamentals, including phytotoxicity testing, were between 4 and 30 fl oz/100 gal. Efficacy has been demonstrated in research trials across the country with universities and through the IR-4 program. Target pathogens include leaf spots such as Alternaria and Myrothecium, downy mildews including impatiens downy mildew, Botrytis and other flower and petal blights, Powdery mildew,

Rust, Scab and others. Excellent control has been achieved with the rates tested; many pathogens are controlled with 8 to 10 fl oz/100gal. This combination has also been shown in field trials to be an effective resistance management tool when used in rotation. Crop safety at a 3-4X safety factor has been established in many ornamentals, and a plant list will be included on the label. Efficacy studies were conducted on many ornamentals including Azalea, Boxwood, Dahlia, Daylily, Dieffenbachia, Geranium, Gerbera Daisy, Impatiens, Rose, Spathiphyllum, Stock, Tulip, and Zinnia. Registration is expected first quarter of 2016.

BIOLOGY, ECOLOGY, TAXONOMY, AND EVOLUTION

VARIATION OF MANDIBULAR GLAND VOLATILES IN THE APTEROSTIGMA PILOSUM SPECIES GROUP.

C.T. Hogan*, T.T. Jones, M. Zhukova, J. Sosa-Calvo, and R.M. Adams, Virginia Military Institute, Roanoke, VA (209)

ABSTRACT

Fungus-farming ants faithfully cultivate five phylogenetic groups of fungi that they use as their primary source of food. Members of the *Apterostigma pilosum* species group cultivate a basidiomycete fungus belonging to the coral-mushroom family Pterulaceae that is distinct from all other cultivars grown by farming ants. A chemical analysis using gas chromatography/mass spectroscopy of the mandibular gland volatiles of two species – *A. dentigerum* and *A. manni* – revealed the presence of an extraordinary diversity of natural products including a number of compounds new to Arthropoda. These results support the phylogenetic divergence of the two species sampled from the *A. pilosum* group but also reveal that there is remarkable convergence and divergence of alarm compounds across the fungus-growing ants.

USING MITOGENOMES IN PHYLODYNAMIC STUDIES OF *IXODES SCAPULARIS*.

C.E. Khatchikian*, M. Prusinski, M. Stone, L.J. Meehan, P. Backenson, I. Wang, M.Z. Levy, and D. Brisson, University of Pennsylvania, Philadelphia, PA (211)

ABSTRACT

The blacklegged tick, *Ixodes scapularis*, is the most important vector of human diseases in Northeastern and Midwestern United States. Populations of blacklegged tick have established and flourished in areas of North America previously thought to be devoid of this species. Moreover, this increase in the geographic range of tick populations parallels the expansion in the human incidence of Lyme disease and other vector-borne diseases vectored by *I. scapularis* over the last decades. Recently, we showed the importance of population growth and progressive and long-distance migration for the range expansion of *I. scapularis*, using serial coalescent and phylodynamic modeling approaches. These approaches were based on multiple mtDNA loci, taking advantage of mitochondrial high mutation rates and single coalesce. However, a few loci could inherent lack the necessary temporal and spatial resolution to fully resolve the demographic histories, direction, rate, and timing of migration at all ecologically relevant scales. In this work, onsidering the number of variable SNPs available across the entire mitochondrial genome, we propose the use of mitogenomes (complete mitochondrial genomes) for these studies. We present a mitochondrial amplification method that reaches more than 30x coverage across 96 samples on a single Illumina MiSeq run (~1.5 million 150-bp reads per run) across the entire tick mtDNA. The preparation time, amplification and sequencing cost suggest that this approach is feasible and the cost is comparable to the conventional multi loci sequence approaches using Sanger-based technology.

GENOMIC AND MOLECULAR APPROACHES TO INVESTIGATE GLYPHOSATE RESISTANCE â€“ GIANT RAGWEED CASE STUDY. B. Schulz*, University of Maryland, College Park, MD (212)

ABSTRACT

Giant ragweed *(Ambrosia trifida L.)* is a competitive annual weed prominent in agriculture fields and cultivated areas of the Midwestern U.S. It is one of the most problematic weeds in corn and soybean production systems. The introduction of glyphosate resistant agronomic crops (“Roundup®-ready”) in 1996 provided an effective tool to manage giant ragweed. The physiological mechanism of glyphosate’s herbicide effect is inhibition of EPSP synthase, a key enzyme in the shikimate pathway, which produces aromatic amino acids and many other plant signaling compounds. The use of glyphosate based herbicides drastically increased after 1996 due to overreliance and repeated use of glyphosate for weed control in glyphosate resistant crops. These use patterns resulted in tremendous selection pressure for glyphosate resistant giant ragweed (GRGR). GRGR has been reported in several states of the U.S and in Canada. We are investigating the mechanism(s) of glyphosate resistance in Midwestern GRGR biotypes. The ultimate goal of our project is to discover and isolate genes responsible for glyphosate resistance in GRGR. We hypothesize that the basis of resistance in GRGR biotypes is related to reduced translocation of glyphosate and a rapid response of glyphosate treated leaves in GRGR, which show a hypersensitive-like (HR) reaction to herbicide treatments. This HR results in leaf abscission within a day of treatment. GRGR plants do not die from glyphosate treatments but resume normal-growth from shoot and axillary meristems and reproduce. Glyphosate sensitive plants do not show a comparable HR and transport the herbicide throughout the entire plant, which leads to leaf chlorosis and eventual death of the treated plants after 2-3 weeks, likely due to inhibition of EPSPS.

The two biotypes were compared for rate response to glyphosate at typical field rates of 1X, 2X, 4X and 8 X application rates (1X=0.84kg/ha). We found that GRGR plants survive rates of 6.72 kg/ha. To assess the reaction of sensitive and resistant biotypes to glyphosate on the molecular level we analyzed the total transcriptome of treated and untreated plants after glyphosate application.

GR plants had a unique response when treated with glyphosate, exhibiting initial rapid necrosis of mature leaves within 12 hours of treatment. GR plants do not die from a glyphosate treatment but resume normal-growth from axillary meristems and reproduce. The progression of the response and symptoms resemble a typical hypersensitive response similar to that observed on some plants after pathogen attack. GS plants do not exhibit rapid leaf necrosis but their leaves become chlorotic, then necrotic and plants die over a 2-3 week period. We have identified a list of genes that are differentially expressed between the two biotypes as the first step in identifying genes responsible for the glyphosate resistance observed. Results show that GR plants will persist in our current cropping systems if glyphosate continues to be the main weed control tool.

GROWTH AND SEED PRODUCTION OF GLYPHOSATE-RESISTANT AND -SENSITIVE PALMER AMARANTH IN THE FIELD. C.W. Cahoon*, A.C. York, D.L. Jordan, P.J. Tranel, and M.D. Inman, Virginia Tech, Painter, VA (213)

ABSTRACT

Biotypes of Palmer amaranth expressing resistance to glyphosate have made managing this weed in cotton and other crops challenging. Numerous experiments have been conducted to determine if glyphosate resistance carries a fitness penalty in Palmer amaranth. Research conducted by labs in Colorado in the US and in Australia suggests that there is no fitness penalty associated with glyphosate resistance in this weed. Research was conducted in North Carolina in the field to determine if there is a relationship between EPSPS gene copy number, an indicator of glyphosate resistance, and seed production of Palmer amaranth in presence of cotton. Native populations of Palmer amaranth were evaluated in North Carolina near Clayton and Mount Olive in 2014 by allowing approximately 100 plants at each location to interfere with cotton for the entire season. Late in the season EPSPS gene copy number from each plant was determined. Seed was collected after reaching maturity but before shattering, dried in the greenhouse and cleaned in order to determine total seed weight and subsequently seed number per plant. At Clayton, the population of Palmer amaranth was comprised of 46 male and 47 female plants. Of the male and female plants, 34 and 36 were resistant to glyphosate (EPSPS copy number > 2). All other individuals in the populations were considered susceptible to glyphosate. At Mount Olive, 47 males and 59 females made up the entire field population. Male and female plants resistant to glyphosate numbered 29 and 35, respectively. All other plants were considered glyphosate-susceptible. Average seed production approximated 540,000 and 480,000 at Clayton and Mount Olive, respectively. Differences in seed production existed between resistant and susceptible females at each location. At Clayton, glyphosate-resistant female plants produced approximately 445,000 seed. This was significantly ($p = <0.0001$) less than glyphosate-susceptible females which produced around 630,000 seed. Similarly, at Mount Olive, glyphosate-susceptible females produced less seed than glyphosate-resistant plants ($p = <0.0001$). At this particular location, approximately 310,000 and 630,000 seed were produced by glyphosate- resistant and –susceptible females, respectively. These data suggest a possible fitness penalty due to glyphosate resistance based on total seed production in the field. However, while gene copy number expressed in the female parent was known, pollination occurred from a mixed pool of male parents with respect to glyphosate resistance.

PHYSIOLOGICAL COLOR CHANGE IN RESPONSE TO MATING BEHAVIOR IN *ARGIA APICALIS* (ODONATA: COENAGRIONIDAE). A. Whispell*, Rutgers University, New Brunswick, NJ (216)

ABSTRACT

In Odonata, the gradual development of adult coloration during their teneral period is a well-known phenomenon, and this type of morphological color change occurs slowly and irreversibly. Physiological color change, in contrast, is always rapid and reversible, and has only been studied in a few insect species. While the change exhibited by most odonate species is primarily temperature sensitive, changing to dark-phase coloration (DP) when below a certain temperature threshold, then returning to bright-phase coloration (BP) when the temperature rises above it, this is not always the case. *Argia apicalis* males also change colour, from BP to DP, in response to mating-associated activities. In order to study this color change, we marked and released a total of 859 males over the course of three summers, subsequently resulting in data documenting 117 separate mating events. Within each one-month period of data collection, we found that significantly more males changed from BP to DP over the course of a mating event (Mann-Whitney U-Test, $z = 3.31$, $P < 0.001$), than remained in BP throughout (mean ± SE: DP 13.5 ± 0.56, N=108; BP 1.13 ± 0.10, N=9). Sufficient data was obtained from 26 of these mating events to warrant our examining which mating-related behaviors could be correlated with the greatest overall thoracic darkening – tandem formation, copulation, or oviposition – and the males in all 26 cases exhibited their darkest coloration during oviposition. Since colors are often so important in intraspecific communication, it could be advantageous for *A. apicalis* males to publicize their readiness to mate while in BP, and then reduce their conspicuousness when at their most vulnerable, by transitioning to DP.

NEW INSIGHTS INTO HOST DEFENSE MECHANISMS FOLLOWING ARTIFICIAL INOCULATIONS WITH THE BUTTERNUT CANKER PATHOGEN. D. Rioux*, M. Blais, N. Nadeau-Thibodeau, M. Lagace', P. DesRochers, K. Klimaszewska, and L. Bernier, Natural Resources Canada, Quebec City, QC (217)

ABSTRACT

Ophiognomonia clavigignenti-juglandacearum is considered to be the major pest of butternut and is threatening the tree's survival over its native range. In Canada, butternut was listed as endangered under the *Species at Risk Act* in 2005. While identifying, propagating and testing butternut putatively resistant to butternut canker, we found that a period of dormancy was necessary for the expression of symptoms in most seedlings that were artificially inoculated. Before dormancy and in some seedlings that seemed resistant even after dormancy, defense reactions such as wound closure and compartmentalization were obvious and contributed to greatly limiting pathogen development. Suberized tissues were detected in compartmentalization tissues, but most of them were only part of new periderms formed in the bark around the invaded portions of the stem. Phenolic compounds were commonly observed components of the defensive tissues and at times were intensely formed in a continuous band internally to the necrotic xylem that seems to correspond to a new reaction zone of the compartmentalization model. Tyloses regularly blocked the vessels of the invaded xylem. Following dormancy in most seedlings, the initial defensive barriers were regularly breached by the pathogen and then additional ones were tentatively produced by the host to contain the new lesions. These results indicate that some butternuts could be resistant to this canker and thus should be retained in programs aimed at conserving and restoring viable butternut populations.

ASSESSING INTERNAL DECAY IN TREES NONDESTRUCTIVELY WITH TOMOGRAPHY. R.E. Marra*, N.J. Brazee, and S. Fraver, The Connecticut Agricultural Experiment Station, New Haven, CT (218)

ABSTRACT

Through carbon (C) sequestration, forests play a critical role in mitigating against global warming caused by rising concentrations of atmospheric C. Internal decay of wood in living trees, which can extend over most of a tree's lifespan, is a countervailing metabolic process, resulting in release of sequestered C in the form of greenhouse gases. Despite its importance and near ubiquity, internal decay remains unaccounted for in models of forest C dynamics, an understandable omission given the absence of methods for nondestructively quantifying its rate and extent. Our goal was to develop, test, and validate a methodology for this purpose. We used sonic and electrical resistance tomography (SoT and ERT, respectively) to generate three SoT and ERT cross-sectional images on the lower boles of each of 72 trees (23 *Acer saccharum*; 21 *Betula alleghaniensis*; 28 *Fagus grandifolia*) in Great Mountain Forest, Norfolk, CT. 105 cross-sections corresponding to the tomographs were cut from approximately half the trees, after felling, and then processed for photography, gas chromatography, and dendrochronology. Tomographic variation correlated positively with C density, and accurately predicted presence and size of cavities in 68 percent of the cookies. Tree ages, which varied from 68 to 192 years old, permitted estimation of lifetime rates of decay. We conclude from this work that SoT and ERT constitute a powerful tool in identifying and quantifying internal decay in living trees.

WHICH SPECIES TRAITS ALLOW PLANT SPECIES TO PERSIST IN SOWN GRASS-MARGIN STRIPS. S. Cordeau*, M.R. Ryan, D.A. Bohan, X. Rebound, and B. Chauvel, Cornell University, Ithaca, NY (219)

ABSTRACT

Sown grass-margin strips, which have typically been established to limit pesticide drift and soil erosion, are now also promoted for their positive, ecological impact on floral diversity. We ask: i) whether there are traits or combinations of traits that allow weed species to persist in sown grass strips; and, ii) how does duration of sown grass strips affect weed community composition? We hypothesized that therophyte annual and small weed species will dominate the community in newly sown grass strips (<2 years old) and that competitive and perennial weed species dominate the community in older strips (>5 years old). Floral surveys were performed in 75 sown grass-margin strips in France and each strip was surveyed twice. Weed species were summarized with trait databases into functional groups using Multiple Correspondence Analysis and Hierarchical Ascendant Classification. Functional group trajectories were investigated in sown grass-margin strips that differed in age using a space-for-time substitution. We found that geophyte, competitor and monocotyledon species were more frequent and abundant than therophyte, ruderal and dicotyledon species. Trait combinations found in most functional groups appeared to be tolerant of strip management; however, one functional group consisting of therophyte, ruderals with a continuous reproduction period declined in importance with age of sown grass-margin strips. Our findings have implications for non-crop vegetation management for optimizing diversity and ecosystem services in land enrolled in conservation programs. Specifically, we propose that management should change towards regimes with mowing later in the season and soil scarification by tine harrows if floral diversity is a goal.

PRIORITY RATINGS, BASED ON WEED BIOLOGY, AID IN THE MANAGEMENT OF ANNUAL AND PERENNIAL WEEDS IN CRANBERRY. H. Sandler*, K.M. Ghantous, R. Neron, and J. Deland, UMass Cranberry Station, East Wareham, MA (220)

ABSTRACT

Managing weeds in perennial crops systems present challenges not encountered in annual cropping systems. Perennial weeds are usually well-adapted (e.g., having rhizomes, stolons, tubers), making control very difficult. Managing weeds in cranberries has another level of difficulty since the cranberry vines grow to form a continuous mat across the ground. Weeds are often intermingled with the vines making postemergence herbicide applications problematic due to either crop safety concerns and/or getting enough herbicide to the target plant. Many growers struggle to determine which weeds are the most important to control and on which weeds they should expend their limited time and financial resources.

We initially developed a priority grouping system (1995) to help growers prioritize their weeds. This system categorized weeds into 4 groups based on subjective evaluation of three criteria: ease of control, ability to spread, and impact on yield. During the creation a new weed guide, we revised and expanded the prioritization. The Priority Rating of each weed was determined by considering four criteria: impact, biological form or type, invasive or reproductive capacity, and adaptation to the cranberry habitat. For each weed, a score of 8, 4, 2, or 1 was assigned to each criterion. The impact of the weed on the cranberry plants themselves, 8 = killing or significantly crowding out vines; 4 = reducing vine vigor; 2 = reducing yield; and 1 = little effect. For biological form, 8 = perennial similar to cranberry; 4 = perennial different than cranberry; 2 = perennial or biennial plant; and 1 = annual plant. For reproductive or invasive capacity, 8 = vigorous stolons and rhizomes; 4 = low to moderate production of stolons and rhizomes; 2 = propagation by seed only with many seed; and 1 = propagation by seed only with few seed. Well-adapted plants that are hard to pull received a score of 8; well-adapted plants that are easy to pull scored a 4; 2 = marginally adapted, hard to pull; and 1 =

marginally adapted, easy to pull. The total number of points determined the final Priority Rating: 4 to 7 points (Low), 8 to 15 points (Medium), 16 to 23 points (High), and 24 to 32 points (Very High). By using a 4-tiered ranking system within each of the selected criteria, the revised Priority Ratings distributed the weeds along a wider continuum. The new system created 32 possible scores compared to the previous system of 4 scores. The broader distribution gives growers greater precision for prioritization.

We have ranked 144 weeds in the recently published English-version of the *Identification Guide for Weed in Cranberries* (2015). The priority rankings are denoted for each weed with a corresponding number of squares in the upper right-hand corner of the page; readers can quickly see the Priority Rating when viewing the photographs. Readers can then understand how each weed attained its Priority Rating by viewing the ranking tables located towards the end of the publication. Since soil and environmental conditions can vary among cranberry production areas (includes BC, MA, NJ, OR, QC, WA, WI, and the Atlantic provinces), the rankings can be used as a baseline and adjusted to re-define the Priority Rating of a particular weed for any particular area.

REFLECTIONS ON WEED BIOLOGY IN CRANBERRY. K.M. Ghantous* and H. Sandler, UMass Cranberry Station, East Wareham, MA (221)

ABSTRACT

Managing weeds in cranberry presents many challenges. Cranberry is a longâ€'lived woody perennial, which renders many traditional weed management practices such as tillage and crop rotation unusable. Cranberry is also unlike other perennial crops such as tree fruits and grapes in that cranberry vines are low growing, trailing vines that create a continuous mat of vegetation with no rows or open areas to cultivate around or below the crop. Dodder (*Cuscuta* spp.) and Poverty grass (PG: broomsedge bluestem (*Andropogon virginicus*) and little bluestem (*Schizachyrium scoparium*) are a few of the most problematic weeds in Massachusetts cranberry.

All of the weeds ranked by growers as being most difficult to control are also perennials like the crop system, with the exception of the annual plant, dodder. Dodder is an obligate parasitic plant, having no apparent leaves and thin stems, leaving little surface area for herbicides to contact. These atypical physical characteristics protect the weed from much direct contact with herbicides, in some cases only being exposed after a chemical is translocated through the host plant into the dodder via phloem and xylem connections. Recent observations indicate that dodder biology may be influenced by grower practices. Seeds collected from multiple sites had significant differences in germination rates; the site with the greatest germination rates averaged 37.6 percent and the site with the lowest averaged 12.8 percent ($p \leq 0.05$). This may, in part, explain why control at different sites using the same herbicides is highly variable.

Some changes in farm management practices have occurred gradually over time (e.g., changes in fertility management, drainage and irrigation practices, and other cultural practices). Some changes occurred more rapidly. Two recent dramatic economic downturns in the cranberry market have led to many growers taking additional jobs, leaving less time for labor-intensive practices, and less money to invest in pesticides and other farm inputs. Many growers have also moved away from spotâ€'treating with herbicides and rely mainly on chemicals that can be applied through the irrigation system. All of these changes have contributed to weed population shifts, some of which have been rapid. PG has been present on farms for decades, but in the past five years, populations have increased in occurrence and severity. In a 2010 survey, no growers selected perennial grasses as their most problematic weed, while 64 percent rated PG as one of the four most commonly occurring weeds, and 59 percent rated it one of the most difficult weeds to manage in a 2015 survey. The number of growers rating moss (*Polytrichum commune* and *Sphagnum spp.) as* one of their four most difficult to manage weeds also increased in this same time period. Further research is needed to understand how weed biology and grower practices in cranberry are interconnected.

MEADOW KNAPWEED IN THE NORTHEAST: SHOULD I KNOW THIS SPECIES? C. Marschner*, L.R. Milbrath, J.M. Hunn, J. Biazzo, and A. DiTommaso, Cornell University, Ithaca, NY (222)

ABSTRACT

Meadow knapweed (*Centaurea x moncktonii*) is increasingly prevalent in New York agricultural lands, including pastures, meadows and waste areas. It is a hybrid of black (*C. nigra*) and brown (*C. jacea*) knapweeds that can be similar in appearance to the parental species, and we believe that meadow knapweed may be more prevalent in our region than previously suspected. In New York, its populations appear to be expanding. Very little is known about this hybrid species, including such basic information as its germination characteristics. We are running a series of germination trials to better understand the potential seasonal timing of seedling emergence. Preliminary data indicate that although cold stratification increases the rate of germination, particularly at lower temperatures, it is not necessary for germination to occur. Light increases germinability, but is not absolutely necessary. Germination is most rapid at warmer temperatures (25/15C and 30/20C light/dark treatments). Seeds in the dark treatment at the lowest temperature (15/5C) had minimal germination regardless of stratification. There appear to be interactions between light, temperature and stratification variables. This study will be continued to clarify these interactions, and examine the germination trait variability between populations of this widespread Centaurea hybrid species.

VEGETATION MANAGEMENT AND RESTORATION

DOES SUSTAINED WEED SUPPRESSION IMPROVE ESTABLISHMENT OF WILDLAND PLANTINGS OF SHELTERED SILVER MAPLE, SYCAMORE, OR PIN OAK? A. Gover* and R. Reese, Penn State, University Park, PA (223)

ABSTRACT

One approach for managing exotic annual species with persistent seed banks is reforestation. The rationale is to invest effort in establishment and release for a few growing seasons, in contrast with annual suppression efforts that potentially could extend indefinitely. Riparian sites with infestations of mile-a-minute (*Persicaria perfoliata* (L.) H. Gross, POLPF) at Bald Eagle State Park (BAEA), Howard, PA, and Sinnemahoning State Park (SINN), Austin, PA, were planted to silver maple (*Acer saccharinum* L.), American sycamore (*Platanus occidentalis* L.), and pin oak (*Quercus palustris* Münchh.) at two planting densities (500 or 1,000/ha), with three post-planting maintenance periods (0, 2, or 4 years). The experimental design is a randomized complete block, split-split-plot with planting density as whole plot, maintenance interval as sub-plot, and three plants of each species randomized within each subplot. Seedling trees were planted into augered holes at BAEA on April 12 and 13, 2012, and at SINN on April 30 and May 1, 2013, then enclosed in 1.5 m rigid plastic shelters. Shelter maintenance is uniform across treatments, and weed maintenance periods of 0, 2, or 4 years consisted of annual applications of glyphosate plus sulfometuron plus prodiamine at 3.4 plus 0.053 plus 1.6 kg/ha to 1.8 by 1.8 m squares around each shelter, plus a single spot treatment of POLPF and other exotic target weeds within each sub-plot using glyphosate plus triclopyr at 3.4 plus 1.7 kg/ha. At initial planting and each fall through 2015, tree height was recorded to the highest viable bud. At this early stage of reforestation, planting density was not significant, therefore each planting density whole plot was treated as a replicate (n=6), and the analysis of variance was conducted as a split-plot, with maintenance interval as whole plot, and species as subplot treatment. At BAEA, growth was satisfactory, and growth behaved according to assumption, as the trees with weed suppression grew more than those without, beginning the first season. The 4- and 2-year maintenance regimes produced significantly greater growth than the 0-year treatment for the first and second growing seasons, and the 4-year treatments produced better growth in growing seasons 3 and 4, remaining significantly higher than the 0-year, while the 2-year treatments settled into an intermediate level, significantly different from neither the 0- or 4-year treatments. At SINN, growth was satisfactory, and there were no growth differences between the maintenance treatments during the first three growing seasons. Potential factors contributing to the site difference include the pre-plant herbicide regimen (more extensive at SINN), site soils (more uniform and mesic at SINN), and different plant materials sources for the two sites. Subsequently, we have intitiated experiments examining the interaction of site preparation and follow-up maintenance intervals.

USING A METRONOME TO STANDARDIZE BACKPACK SPRAYER CALIBRATION AND APPLICATION GROUND SPEED.

B.F. McDonnell* and G. Ciabattoni, National Park Service, Bushkill, PA (224)

ABSTRACT

: Backpack sprayers are commonly used in vegetation management. Proper calibration and use of such equipment require the applicator to standardize several variables between equipment calibration and subsequent applications. These factors include application spray pressure, spray tip height, spray pattern overlap of adjacent spray rows, and ground speed. Standardizing ground speed can be difficult, especially when the calibration tests are conducted on flat, level ground, while field applications are often done over uneven and variable terrain. A calibration and application method using a metronome to standardize ground speed is presented. The metronome is used to control stride pace. Benefits include a more consistent rate of application over the target area, and a calibration process that requires very little ground application compared to other methods, such as the 1/128 method. The metronome method requires the applicator to determine the effective spray pattern width, but does not require application to a predetermined shape on the ground. The metronome method does require the user to perform some simple mathematical calculations.

INTRODUCING A WINDOWS PC PROGRAM FOR PESTICIDE APPLICATION CALCULATIONS. B.F. McDonnell*, National Park Service, Bushkill, PA (225)

ABSTRACT

Two of the most important and basic duties of a pesticide applicator are to properly plan for an application, and properly mix the final pesticide mixture to be applied. Ideally, the applicator will know how much of each tank mix partner will be needed for the entire application, and how much of each tank mix partner should be added to each tankful. This can often require tedious and repetitive calculations with a handheld calculator, which is susceptible to user entry error leading to misapplication of pesticides. A Windows® -based application is presented, which may be run on a portable computer, such as a laptop or tablet. The user enters data about the application and products to be applied. The program performs the necessary calculations, and produces a text document of results. The results document repeats the user's entries, which helps limit user entry error, estimates the total weight or volume of each product for the entire application, and displays the proper volume or weight of each product to be added to each tankful.

A HANDHELD PRUNING DEVICE FOR SYSTEMIC INJECTION OF HERBICIDES OR CROP PROTECTION CHEMICALS. S.D. Askew* and T.B. Burch, Virginia Tech, Blacksburg, VA (226)

ABSTRACT

Treating stumps of cut vegetation with pesticides has been shown to either kill the living stump or protect the living stump from infection by pathogenic agents depending on the type of pesticide employed. In research studies, such treatments are applied immediately after cutting or within 1 to 4 hours of cutting with applications made sooner typically performing better than when applied later. For large scale operations, timing of cutting operations and spraying operations are not always optimal. An apparatus that can deliver treatment fluid to cut stems during the cutting process would, in theory, improve effectiveness of the treatment fluid and reduce labor by eliminating need for subsequent spray treatment. A new invention by Thomas Burch applies this concept in a device that can be used to prune desirable plants, such as fruit trees and ornamental shrubs, or cut small weedy vines, shrubs, or trees. An apparatus for applying liquid treatment was configured for attachment to a conventional reciprocating saw equipped with pruning blades. The applicator assembly includes a housing that defines a recess for receiving the vegetation. Entering stems trigger a mechanism that initiates fluid application to the cutting blade in such a way that the cut stump or living stem receives chemical treatment during the cutting process. Studies at Virginia Tech are currently comparing translocation of radio-labeled herbicide into cut stems when herbicide is applied with the pruning device or as a spray applied at various times after cutting. Potential applications of this technology include consumer-level devices that can control unwanted vine or woody vegetation in the home landscape and commercial devices that can control or prevent pathogenic agents in production horticulture crops, citrus crops, or other applicable industries.

CAN REDUCED-DILUTION, BARK-APPLIED MIXTURES OF IMIDACLOPRID PLUS DINOTEFERAN CONTROL HEMLOCK WOOLY ADELGID AND IMPROVE PRODUCTIVITY?

T.R. Mazzone*, A. Gover, R. Reese, D. Schmit, and B. Brantley, Pennsylvania State University, State College, PA (227)

ABSTRACT

The eastern hemlock (*Tsuga canadensis*) is a keystone tree species whose survival is seriously threatened by the exotic insect hemlock wooly adelgid (*Adelges tsugae*, HWA) and elongate hemlock scale (*Fiorinia externa*, EHS). Infestation and loss of canopy eastern hemlock trees results in significant changes in plant communities and riparian ecosystems. As a result, these trees may be worthy of preservation efforts in some forest settings. The eastern hemlock is often protected by PA State Parks or Forests by targeted soil injections of imidacloprid. Injectors are set up to deliver a diluted dose of 15 mL/2.5 cm of stem diameter at breast height (DBH), and used in conjunction with a 3 gallon backpack sprayer. This configuration allows for treatment of 960 inches of DBH per backpack. Trunk sprays of dinotefuran are labeled for use at 78 to 96 mL/2.5 cm DBH. This required volume results in less efficiency in treating trees and loss of productivity. As a result of the current standards, insecticide delivery is often difficult in remote areas, due to large carrier volumes needed for trunk sprays and rocks inhibiting use of soil injectors. Imidacloprid, the soil injection standard material, has lagged activity but a long residual component. Dinotefuran, the trunk spray standard, is a fast-acting, short residual activity material. This study observed the insecticidal effects of imidacloprid-dinotefuran tank mixed basal spray, at near-standard volume (24 mL/cm DBH) and low (8 mL/cm DBH) carrier volume, as compared to imidacloprid soil injection or dinotefuran basal spray at standard carrier volumes. The utility of reducing carrier volume and combining active ingredients could help ecological preservation efforts protect more trees, with less return trips and provide both short and long term control of the HWA. Insecticidal treatments were applied at Caledonia State Park, Fayetteville, PA in October 12, 2010. Pre (October 2010) and post (Winter 2011, 12, 13) treatment insect counts were conducted by surveying 10 random, 25-cm terminal shoots per tree. Insects were counted from 0 to 10, 10 being the maximum. Over four years, insecticidal response of EHS populations was not affected compared to the control. In 2011, HWA populations declined in response to dinotefuran (range, 0.7-1.8 insects), when compared to the untreated control (6.0 insects). HWA populations did not decline in response to imidacloprid in 2011. All treatments showed increased insecticidal activity (range, 2.1-2.9 insects) in 2012, compared to the control (6.5 insects). 2013 HWA response to dinotefuran treatments (8.8-9.5 insects) was similar to the control (10.0 insects). Imidacloprid treatment had a reduced HWA count (8.0 insects). Tank mixing did not appear to offer advantageous insecticidal effect. These results suggest that low carrier volumes of dinotefuran can be used for suppression of HWA. Low dinotefuran carrier volumes can provide applicators a means to treat more trees per tank and increased preservation of eastern hemlock canopy.

MANAGEMENT OF HYDRILLA IN A FREE FLOWING RIVER. R.J. Richardson*, S. Auell, and S. Hoyle, North Carolina State University, Raleigh, NC (228)

ABSTRACT

Hydrilla (*Hydrilla verticillata*) is an invasive aquatic weed that has been spreading throughout North Carolina's lakes and reservoirs since it was first discovered in 1980. It is now invading increasingly dynamic and high biodiversity systems such as rivers and natural lakes. One recent site of invasion is the Eno River system in the Piedmont region of the state. The Eno is is a tributary of the Neuse River, and is home to several rare species including the panhandle pebblesnail (*Somatogyrus virginicus*). It also serves as a significant source water for Falls Lake, the drinking water reservoir for the City of Raleigh, NC and several surrounding areas. In 2015, an aquatic herbicide treatment with fluridone (Sonar Genesis) was conducted in the Eno River, marking the first metered herbicide treatment of hydrilla within a riverine system in the state. We evaluated the herbicide treatment impacts to selected target and non-target aquatic species. Efforts included quantitative sampling of *H. verticillata, S. virginicus*, and *Podostemum ceratophyllum* (the native vegetation and habitat of *S. virginicus*) at seven spatially separated sites along the Eno River. Biweekly vegetation monitoring and monthly snail sampling began in late May, two weeks before treatment, and continued through December. *H. verticillata* shoot lengths were significantly reduced during treatment from an average of 23.4 cm to 10.6 cm. Average density of *S. virginicus* was significantly different among sites, ranging from 5,537 snails/m^2 to 1,782.4 snails/m^2 . Monthly snail density averaged among all sites differed over the course of the sampling season, with lower densities found in October and December. Average monthly snail densities during treatment months did not differ significantly. *P. ceratophyllum* densities differed between treated and untreated sites with means of 13,736 and 10,682 stems/m^2 , respectively. Overall, fluridone effectively reduced hydrilla density within the treated area with no apparent negative impact to the studied non-target species.

HYDRILLA IN THE NORTHEAST: MY PERSPECTIVE ON RESEARCH, MANAGEMENT, AND THE FEASIBILITY OF ERADICATION. R.J. Richardson* and M. Netherland, North Carolina State University, Raleigh, NC (229)

ABSTRACT

Hydrilla (*Hydrilla verticillata*) is a highly invasive submersed aquatic plant. Two biotypes are present in the US, a female triploid dioecious and a triploid monoecious with these typically segregating into southern and northern regions, respectively. There are at least three major biotypes worldwide and cryptic speciation is a possibility. Monoecious hydrilla is a prolific turion producer and these overwintering propagules are the primary challenge to management. New detections of monoecious hydrilla continue in the Northeast in waterbodies such as Croton River, NY, and Pymatuning Reservoir, OH and PA. While hydrilla does continue to expand in cooler regions of the US, growth of the plant is primarily limited by water temperature. Monoecious hydrilla can sprout from subterranean turions and form new turions in as little as five weeks with water temperature of 16 C or greater. Observations of turion sprouting from NY indicate that a greater percentage of turions sprout annually, likely stimulated by cooler winter temperatures. Greater sprouting percentages also mean that the turion bank can be reduced more quickly by management. While fewer management options are present in the Northeast US as compared to the Southeast US (e.g. triploid grass carp), management in the Northeast may reach functional eradication more quickly. Due to the high diversity of submersed aquatic plant species in glacial lakes, better integration of vegetation surveys with management should be developed to allow management of hydrilla while protecting desirable species.

SUPPRESSION OF JAPANESE HOPS WITH EARLY-SEASON HERBICIDE APPLICATION. A. Gover*, The Pennsylvania State University, University Park, PA (230)

ABSTRACT

Fourteen herbicide treatments and an untreated control were initiated April 1, 2010 to suppress Japanese hops (*Humulus japonica* Sieb. & Zucc., HUMJA) in a riparian setting in The Laurels Preserve in Unionville, PA. The primary objective was to identify alternatives to sulfometuron for pre- or early postemergent applications. Treatments (in kg/ha) included sulfometuron at 0.053; pendimethalin at 4.5 alone or combined with imazapic at 0.017 or 0.035, triclopyr at 0.42, or metsulfuron at 0.021; prodiamine at 1.6 alone, or combined with imazapic at 0.017, triclopyr at 0.42, or metsulfuron at 0.021; imazapic alone at 0.10 or 0.21; or flumioxazin at 0.29 or 0.43. Treatments were applied to 1.8 by 4.6 m plots arranged in a randomized complete block with three replications. Predominant vegetation was reed canarygrass (*Phalaris arundinacea* L., TYPAR) quackgrass (*Elymus repens* (L.) Gould), and poison hemlock (*Conium maculatum* L.). Monthly assessments of HUMJA and collateral vegetation response were taken May 4, June 1, and June 29, 2010.

On June 29, the untreated plots averaged 98 percent total vegetative, 4 percent HUMJA, 43 percent TYPAR, and 13 percent quackgrass cover. The treatments featuring pendimethalin, prodiamine, or flumioxazin averaged 93 percent or greater total cover, and 4 percent or less HUMJA cover, and in general had little effect against TYPAR. Sulfometuron and imazapic were very active against TYPAR, with different results. Sulfometuron-treated plots averaged 38 percent total cover, and 0 percent HUMJA, and 2 percent TYPAR; while imazapic released HUMJA, averaging 78 and 95 percent HUMJA cover at the 0.10 and 0.21 kg/ha rates, respectively.

BAYER VEGETATION MANAGEMENT – NEW VENTURES. David Spak and J. Belcher, Bayer Environmental Science, Research Triangle Park, NC (231)

ABSTRACT

Vegetation management is an ever increasing challenge across the globe. Negative impacts of undesirable vegetation include reduction of land and water resources, destruction of wildlife habitat, deterioration of infrastructure, and impairment of human safety. Late in 2014, Bayer CropScience acquired the Dupont Crop Protection Land Management business which included a portfolio of herbicides for weed control in vegetation management, forestry, and range and pasture markets in North America, Australia, and New Zealand.

Bayer has committed resources toward assisting land managers, applicators, landowners and utility foresters with a comprehensive weed control portfolio, comprising innovative and established herbicides marketed under brand names such as Esplanade™, Method™, Perspective™, Escort™ and Velpar™. A review of the Bayer Vegetation Management business, product pipeline, and stewardship organization will be presented.

FRUIT CROPS

ZEUS PRIME XC: A NEW PRE-EMERGENT HERBICIDE FOR GRAPES, BRAMBLES, AND POME FRUIT. J. Reed*, D. Doohan, B.H. Zandstra, and R. Crassweller, FMC, Philadelphia, PA (232)

ABSTRACT

Zeus Prime XC is a 3.5 lb/gallon soluble concentrate liquid formulation containing 3.15 lbs sulfentrazone and 0.35 lbs carfentrazone. Zeus Prime XC is labeled for weed pre-emergence use with some weed postemergence burndown utility in blueberries, bushberries, caneberries, grapes and apples. Zeus Prime XC may be applied safely provide control of such weed species as herbicide resistant and susceptible palmer pigweed, other pigweed species, common lambsquarters, velvetleaf, Pennsylvania smartweed, black and Eastern black nightshade, specific grasses and yellow nutsedge. Zeus Prime XC use rates range from 10 oz product, 0.273 lbs ai/A (0.306 kg/HA) to 15.2 oz product, 0.415 lbs ai/A (0.466 kg/HA) on most soils according soil texture. Zeus Prime XC is an excellent PPO pre-mix herbicide alternative to control glyphosate and ALS herbicide resistant weeds for fruit producers.

DOES DIRECTED APPLICATION OF GLYPHOSATE INJURE APPLE TREES IN ONE YEAR? R.S. Chandran*, West Virginia University, Morgantown, WV (233)

ABSTRACT

A one-year field experiment was conducted in West Virginia, USA, to determine the effect of formulations, timings, and application rates of glyphosate on phytotoxicity to established 'Macoun' and 'Golden Delicious' apple trees up to one year after treatment. A terrestrial formulation of glyphosate with built-in surfactants (Roundup WeatherMax) and an aquatic formulation of the same without surfactants (Shore-Klear) were applied at rates of 2.24 and 4.48 kg ae/Ha, during June, August, and October. Herbicide spray was intentionally allowed to come into direct contact with 15 to 20 cm of the tree trunk above the ground level. Treatments were arranged factorially with a single tree serving as a replicate and each treatment replicated thrice. No interactions were detected between any variables for 'Macoun' apple. One main-effect was detected (application rate) where the higher application rate resulted in reduced levels of trunk growth at a height of 30 cm above ground level, compared to that by low application rate. In 'Golden Delicious' apple, there was an interaction between timing and formulation, however no trends could be ascertained due to variability. No visual symptoms of phytotoxicity were observed for any of the treatments on either variety of apple trees up to one year after treatment.

DIFFERENCES IN THE FREEZING TOLERANCE (LT_{50}) OF SIX CRANBERRY CULTIVARS GROWN IN MASSACHUSETTS. F. Ndlovu*, P. Jeranyama, and C.J. DeMoranville, University of Massachusetts Amherst/ UMass Cranberry Station, Wareham, MA (236)

ABSTRACT

Freezing injury is a major abiotic stress affecting cranberry (*Vaccinium macrocarpon* Ait.). While changes in bud hardiness during spring deacclimation are well known, bud acclimation in the fall is not fully understood. An experiment was designed to assess the timing of acclimation and degree of cold hardiness for terminal buds of six cranberry cultivars collected from a cranberry field in East Wareham, Massachusetts. Samples were taken from mid-September through mid-December, 2014. Cranberry uprights with terminal buds were exposed to a series of freezing temperatures and then evaluated for visible damage. The temperature at which 50 percent of the buds are damaged (LT_{50}) across treatments was determined using the Gompertz function. The LT_{50} temperatures declined with each successive freeze test, indicating increased freezing tolerance with declining fall temperatures. During the first week of field sampling (mid-September), cranberry cultivars had LT50 ranging from -12.9 C in 'Early Black' to -10.1 C in 'Howes'. The exception was during the first week of October when the field was flooded for harvest; during that period cold hardiness ceased but resumed its downward progression the following week. Cultivars differed in both hardiness progression and degree. 'Crimson Queen' and 'Demoranville' took longer than the rest of the cultivars to recover as they needed more than one week to resume the acquisition of cold hardiness. The buds reached maximum hardiness by 19 November 2014 ranging between -27.4 C in 'Early Black' and -26.0 C in 'Howes'. These results have potential practical implications in determining whether or not frost management to protect buds should be implemented in the fall, particularly for early ripening cultivars.

AMBROSIA BEETLE (*XYLOSANDRUS GERMANUS*) INFESTATIONS AND MANAGEMENT TRIALS IN HIGH-DENSITY APPLE ORCHARDS. A.M. Agnello*, D. Breth, and J. Vandenberg, Cornell University, Geneva, NY (237)

ABSTRACT

Black stem borer (BSB), *Xylosandrus germanus*, an ambrosia beetle established in the US for over 80 years, is causing increasing damage in NY and other northeastern states. Normally a pest of hardwood trees, BSB has been documented in dozens of NY apple orchards since 2013, causing tree death and decline in mostly young dwarf plantings of trees ranging from 5–25 cm in trunk diameter. Although these borers are known to attack physiologically stressed trees (e.g., from flood, drought, or cold injury) that are emitting ethanol volatiles, they also have been reported to attack apparently healthy trees. We used ethanol traps to document their occurrence and timing in over 50 NY orchards, and also tested different trunk insecticide sprays for BSB control in waterlogged potted apple trees placed in the orchard to stress them enough to produce ethanol. Trunk and tree damage was assessed among the different treatments; overall, there was a reduction in trunk infestation holes and galleries in the treated tress compared with untreated checks, but results were variable between the two commercial orchard sites. In one site, chlorpyrifos (Lorsban) handgun treatment resulted in lower damage than using the pyrethroids lambda-cyhalothrin (Warrior) or gamma-cyhalothrin (Declare). At the other site, all handgun treatments resulted in generally lower damage than in the checks, but there was no separation among the products tested.

SPORE TRANSPORT IN A VINEYARD. F.J. Ferrandino*, CAES, New Haven, CT (238)

ABSTRACT

The spread of disease caused by pathogens which are disseminated via airborne spores is critically dependent on the details of air flow within and immediately above the plant canopy. For field crops having a relatively homogenous canopy structure the details of the wind field has been well studied. Trained trellised grapes, however, have a unique "elevated hedgerow" structure. The foliage along a row is limited to a height of between 1 and 2 meters and is trained using catch wires to extend less than half a meter perpendicular to the row. Cross-row wind is characterized by a blow-through under canopy air flow and intermittent circulatory eddies between rows. Down-row wind flow is very similar to the wind field above bare ground with some enhanced turbulence. Preliminary measurements using six 3D Sonic anemometers are presented to illustrate the various air flow patterns. The impact of the observed wind fields on the spread of grape diseases discussed.

EVALUATION OF FALL AND SPRING APPLICATIONS OF HERBICIDES TARGETING RESISTANT WEEDS IN WILD BLUEBERRY FIELDS. D.E. Yarborough* and J.L. Cote, University of Maine, Orono, ME (239)

ABSTRACT

Several herbicides, both registered and unregistered, are currently under review for use on wild blueberry. Indaziflam (6.5 oz/a), flumioxazin (12 oz/a), halosulfuron (1 oz/a) and isoxaben (1.33 lb/a) are pre-emergence herbicides, while rimsulfuron (4 oz/a) may be used pre- or post-emergence. We treated an established weed, red sorrel (*Rumex acetosella*), and a newly emerging weed, horseweed (*Conyza canadensis*). Red sorrel is resistant to several herbicides, competes with blueberry, and hinders harvest. The treatments were applied to red sorrel on ten 1-m^2 plots on 14 May 2015 and five of the plots were treated by the grower with hexazinone (0.4 gal/a). Wild blueberry and red sorrel cover and phytotoxicity, as well as percent cover broadleaf weeds and grasses, were evaluated in June and September. The treatments were applied to horseweed on ten 1-m^2 plots on 11 November 2014 except flumioxazin which was applied on 11/26; five plots were also treated by the grower with hexazinone (6.6 pt/a) and diuron (1.6 qt/a, 5/12/15), and mesotrione (3 oz/a, 6/16/15). Effects on wild blueberry cover and phytotoxicity, horseweed and broadleaf weed and grass cover were evaluated in June and July. Horseweed phytotoxicity was not rated as the plants were either dead or unaffected. Data were analyzed using Tukey's tests to determine significant differences ($\alpha=0.05$). There were no significant differences in wild blueberry cover in the red sorrel trial at either evaluation. Phytotoxicity was initially highest with halosulfuron and was significantly higher than all other treatments except rimsulfuron alone, with no differences in phytotoxicity by September. Although there were initially no differences in red sorrel cover; phytotoxicity was highest in the flumioxazin-grower treatment and was significantly higher compared to the check and other herbicides alone, except for Chateau alone. In September, red sorrel cover was almost eliminated in the grower-treatments. Weed cover was 0-14 percent cover overall, and there were no differences in broadleaf or grass cover. Wild blueberry cover in the horseweed trial was low because the horseweed occurred in bare spots at this site. There were no significant differences in blueberry cover and no phytotoxicity for the treatments alone in June, but phytotoxicity occurred in the flumioxazin treatment in July. The grower-treatments resulted in minor phytotoxicity in June in the rimsulfuron treatment, which was only significantly higher than flumioxazin treatment. However, there was significant phytotoxicity in the grower-treatments in July from the grower's mesotrione post-emergence treatment. At the June evaluation, horseweed cover was significantly lower in the rimsulfuron treatment compared to the other treatments alone (except halosulfuron); by July, rimsulfuron remained lowest but was no longer significantly different. There was no horseweed in the grower-treatment plots at either evaluation. Broadleaf weed and grass cover were low in this trial as well with no differences in broadleaf weed cover at either evaluation. Grass cover was highest in June on the halosulfuron treatment and was significantly higher than flumioxazin and all grower-treatments. In July, grass cover in the halosulfuron treatment almost doubled and was significantly higher than the grower-treatments, plus indaziflam and flumioxazin alone. The addition of indaziflam and flumioxazin improved the effectiveness of red sorrel control when combined with hexazinone and should be evaluated further. Also, fall timing applications of indaziflam and flumioxazin in the prune year should be evaluated to prevent crop year growth of red sorrel in the crop year. Horseweed was not resistant to the mixture of labeled herbicides used by the grower, so it is best controlled with registered herbicides as spring pre-emergence applications.

SOIL AND ENVIRONMENTAL QUALITY

ISOLATING ALLELOCHEMICALS FROM SOIL MICROBIOTA FOR WEED CONTROL. J.T. Kao-Kniffin*, Cornell University, Ithaca, NY (242)

ABSTRACT

Many types of soil bacteria produce secondary metabolites that could be useful to weed management, but isolation of the compounds is challenging. We describe a method to isolate these metabolites through activity-based screening of soil metagenomic libraries. The method allows for the isolation of small molecules produced in vector-host expression systems containing large-insert DNA fragments extracted from the target plant rhizospheres. Allelopathic activities derived from selected clones were screened against a series of controls. Nonmetric multidimensional scaling (NMS) showed similar effects of the set of controls on *Lactuca sativa* growth while *Poa annua* had a broader range of growth responses. Methanol extracts from clones indicating activity showed distinct patterns in grass seedling growth from the empty vector control, but the same extracts showed no effect on *L. sativa.* The results indicate that the metagenomics method and bioassay screen of clone extracts are tools that can be used for initial determination of allelopathic activity from non-cultured soil microbiota.

EVALUATION OF PLANTING DENSITY ON SOIL COMPACTION. S. Yergeau*, Rutgers Cooperative Extension (Ocean & Atlantic Counties), Toms River, NJ (243)

ABSTRACT

Increased compaction of soils, especially during construction, crop planting, or installing and maintaining residential landscapes, accounts for a portion of hydrological alterations to lands in New Jersey. Compacted soils act like impervious surfaces and can hinder water infiltration through underlying soil, creating runoff with greater velocities, larger volumes, and increased flooding. While many methods for managing compaction are available, one option that has not received much evaluation is the use of vegetation after soil compaction has occurred. As plant roots grow through soils, they have the potential to create additional pore spaces to facilitate water infiltration and decrease compaction. The objective of this study is to assess the impact increased planting density of a compaction-tolerant plant species has on compacted soils under unmowed conditions. Three test beds with four different densities of switchgrass (*Panicum virgatum*; at 0, 9, 25, and 49 plants per square meter) were created and monitored for soil compaction and soil water content prior to planting and on a monthly basis during the growing season (May – September 2015). Soil compaction was measured using a handheld static cone penetrometer and reported as the depth reached for critical compaction, defined as soil resistance of 300 pounds per square inch. Immediately after planting, compaction levels in all plots were not significantly different from each other, regardless of the presence or absence of vegetation, but did show an improvement on compaction when compared to the previous condition of mowed turf. Results indicate that plots with vegetation had compaction levels that were significantly different than plots without vegetation at the end of the growing season. A strong relationship between soil water content and compaction was observed during the study period. Future work will continue the project into additional growing seasons and additional research plots, and use other methods to measure compaction (i.e., ground penetrating radar) for a more accurate assessment of this management strategy.

CREATING ACIDIC COMPOST USING ELEMENTAL SULFUR. M. Taylor* and M. Nndanduleni, Longwood Gardens, Kennett Square, PA (244)

ABSTRACT

Traditional compost typically has a high pH range of 7 to 8.5. Continual application of this material can cause an increase in soil pH. Furthermore, if high pH is an issue, alkaline soil amendments should not be used. Two experiments were conducted to determine the amount of elemental sulfur required to create acidic compost. Experiment one was conducted in containers in a greenhouse. Pelleted S (90 percent S) was blended at 6 rates (0, 0.9, 1.8, 3.5, 5.3 and 7.1 g per L) with finished compost created from wood chips, manure, green plant material and food waste. Containers had a volume of 2.5 L and were watered once a week. Compost pH was measured once per week over 150 days and there were minimal treatment differences for the first 30 days. Compost pH then began to decrease in all treatments receiving more than 0.9 grams of sulfur and this decrease continue for 60 days. At 90 days, the pH ranged from 5.5 to 8.2 and increased with decreasing S rates. In experiment two, pelleted S (90 percent S) was blended at 4 rates (0, 2.3, 4.6, and 9.2 g per L) with compost feedstocks on the front end of the composting process. Each treatment consisted of approximately 10 cubic meters of feedstock material. At the end of the 3 month composting process, the respective compost pH was 6.9, 6.0, 5.2 and 2.9. Results indicate that blending sulfur with compost is an effective way to lower pH. If sulfur can be blended with the feedstocks rather than the finished compost, reaction time to lower pH can be greatly reduced or eliminated.

SURVEYING BIOLOGICAL SOIL HEALTH WITH THE SOLVITA CO_2 RESPIRATION SYSTEM. W.J. Sciarappa*, S. Murphy, R. Barresi, and V. Quinn, Rutgers University, Freehold, NJ (245)

ABSTRACT

Measurement of soil respiration CO_2 derived from micro and macro organisms is a potentially important tool to predict availability of nitrogen, the essential element not measured in routine soil tests. Four hundred soil test kits and supplies were utilized including a digital color reader and gel paddles from the Solvita® company - a Division of Woods End Laboratories, Inc. In 2013 and 2014, 18 representative soil sites were selected in Monmouth County, NJ; primarily farms having sandy loam soils with pH values ranging from approximately 5.1 to 6.3 and typical organic matter from 1.0 to 2.0 percent. Seventy-four crop fields were GPS/GIS mapped and sampled at the same location in spring, summer and fall. Six crop categories of local horticulture were annual crops, golf greens, perennial fairways, residential lawns, perennial grass crops and organic blueberry. The first year average for all color values on a 0 to 5 scale was 2.41 with a standard deviation of 0.69 while the second year average was 2.50 with a standard deviation of 0.51. Similarly, the CO_2 respiration values showed no statistical difference between years with values of 16.36 and 13.96 CO_2-C, respectively. Individual soil health results were characterized as marginal to moderate in less fertile areas and medium to ideal in more fertile areas. Soils growing perennial crops as organic blueberry, equine pasture, bio-energy grass or residential lawns reached respective values of 40.1, 23.9, 22.1 and 21.6 CO_2 -C ppm. These values were significantly higher compared to farm soils with annual crops as vegetables and field corn reaching 14.1 and golf course greens 10.6 CO_2-C ppm. Measures of respiratory soil CO_2 serve as a potential indicator of an unmeasured nutrient contribution in sandy loam soils. With the estimated correlation of CO_2 carbon flux to predicted nitrogen contribution, an additional N production predicted for annual crops ranged from about 5 to 15 lbs. per acre annually and 20 to 35 lbs. per acre for perennial crops that could be credited towards total crop needs. These consistent, site specific results in central New Jersey over two years show promising implications for agronomy and horticulture. Farmers, advisors, extension agents and agricultural specialists may be able to add a new assessment method into their technical toolbox.

WHOLE FARM NUTRIENT MASS BALANCES FOR IMPROVEMENTS IN AGRICULTURE AND ENVIRONMENTAL MANAGEMENT OF DAIRY FARMS. Q.M. Ketterings*, S. Cela, M. Soberon, C. Rasmussen, S. Crittenden, and K. Czymmek, Corrnell University, Ithaca, NY (246)

ABSTRACT

Sustainable solutions for agriculture and environmental management on dairy farms require improved nutrient use efficiencies across the entire farm, both for the animals and the cropland. However, when it comes to whole farm nutrient management, nutrient cycling can be very complex and management tools available to farmers often focus on one aspect of management only (e.g. milk urea nitrogen to evaluate crude protein ration management; corn stalk nitrate test to evaluate nitrogen management for corn, etc.). In an adaptive approach for whole farm nutrient management, records are kept in such a way that one can assess the nutrient status of the whole farm, pinpoint the areas where improvements can be made, and then track the progress of those improvements from year to year. A whole-farm nutrient mass balance (NMB) assessment can help farmers and farm advisors do this effectively and efficiently. A NMB is the difference between the amounts of N, P, and K imported onto dairy farms as feed, fertilizer, animals, and bedding, and exported via milk, animals, crops, and manure. We can express a NMB per tillable acre to indicate the potential for recycling nutrients in the land base, an environmental indicator, or per cwt milk, a milk production efficiency indicator. Large positive NMBs per acre suggest high risk of nutrient losses to the environment, while large positive NMBs per cwt reflect low nutrient use efficiencies, and potential economic loss for the farm as well. Negative NMBs (resulting from exports exceeding imports) reflect mining of soil P and K resources, and will eventually reduce crop yields. Annual NMB assessments give farmers a chance to compare the farm against peers in the same milk production group, and to evaluate the impact of management changes on nutrient use efficiency and production. Recently, research has led to the identification of the *optimum operational zone* for nitrogen, phosphorus and potassium. Farms that manage these nutrients in the optimum operational zone recycle nutrients on their land base and produce milk efficiently. Here we present the findings of whole farm balance assessment that resulted in benchmarks for the industy and an adapative management tool that has shows to be effective in identifying opportunities for improvement and in monitoring of impact in both New York State and the Upper Susquehanna Watershed, headwaters of the Chesapeake Bay.

SOIL TEST POTASSIUM BUILDUP AND DRAWDOWN INÂ RESPONSE TO MANURE, COMPOST AND POTASSIUM APPLICATION IN A CORN-ALFALFA ROTATION. A. Sadeghpour*, Q.M. Ketterings, G.S. Godwin, and K. Czymmek, Cornell University, Ithaca, NY (247)

ABSTRACT

Application of manure to meet N needs for corn (*Zea mays* L.) in a corn-alfalfa *(Medicago sativa* L.) rotation, supplies large amounts of potassium (K) during the corn years, typically increasing soil test K (STK) levels and often providing excellent K levels for the first few years of a new alfalfa stand. However, heavy utilization of K by alfalfa result in STK decline. Our primary objective was to evaluate whether K provided by manure application during the corn rotation was sufficient for five years of alfalfa stand. A corn and alfalfa rotation with five years for each crop was initiated in 2001 to compare the impact of N- versus P-based manure (tillage incorporation) and compost management during the corn years on (i) alfalfa yield and (ii) STK trends. Treatments during the corn years were two rates of composted dairy manure (46 and 74 Mg ha^{-1}; averaged over five years) and two rates of liquid dairy manure (68 and 196 kL ha^{-1}; averaged over five years) versus inorganic N as the main N source in the control plots. After five years, alfalfa was established with 90 kg ha^{-1} 0-25-25 ($N-P_2O_5-K_2O$). A K trial was implemented during the alfalfa years using five annual K application rates (0, 93, 186, 282, and 375 kg K_2O ha^{-1}) applied at green-up. The 282 kg K_2O ha^{-1} rate was the estimated crop removal rate for the site, assuming an average crop removal of about 64 kg K_2O ha^{-1} per Mg of DM (2.384 percent K_2O) and an estimated annual yield average of about 9.8 Mg DM ha^{-1} for the Kendaia/Lima soil at the farm. Alfalfa was harvested in a 2, 3, or 4 cut system depending on the weather. Manure and compost additions in the corn years benefited the alfalfa yields in the rotation. Alfalfa yield did not change with a shift from N-based to P-based compost but reduced with shifting from N-based to P-based manure management. Addition of manure and compost increased STK levels by two and three times in N-based and P-based manure and compost treatments during the corn years and declined during the alfalfa year from 135 mg kg^{-1} (averaged over N-based manure and compost) and 96 mg kg^{-1} (averaged over P-based manure and compost) to 75 and 65 mg kg^{-1}, respectively still higher than the initial STK level (47 mg kg^{-1}). Addition of K to alfalfa plots that did not have a compost or manure history (soil test K classified as medium) increased STK levels without improving the yield with the exception of 2008, an exceptionally good growing season, where K addition did increase yield. These data suggest that alfalfa yield benefits from application of compost and manure during the corn years and STK levels can be managed with no addition of K in a corn-alfalfa rotation system.

CANOPY REFLECTANCE MEASUREMENTS TO PREDICT CORN YIELD IN RESPONSE TO NITROGEN FERTILIZATION MANAGEMENT.

A.C. Tagarakis*, Q.M. Ketterings, A. Sadeghpour, G.S. Godwin, I.B. Cornell, and K. Czymmek, Cornell University, Ithaca, NY (248)

ABSTRACT

Proximal (active) sensing has been increasingly used to provide information about the canopy properties in a large variety of crops. We conducted a multi-layer experiment to determine the optimal time of sensing to predict corn (*Zea mays* L.) responsiveness to N fertilization and corn yield using Normalized Difference Vegetation Index (NDVI). We tested three sources of N (liquid dairy manure, composted separated dairy solids, and inorganic fertilizer) and three application methods (broadcast pre-plant, starter plus sidedress, and manure or compost at two rates, with and without additional sidedress N). The trial was conducted at Musgrave Research Farm in Aurora, NY on a calcareous Lima Silt-loam soil. The main experimental design was a randomized complete block design with five blocks and fourteen treatments. Treatments included manure and compost surface-applied without incorporation at rates to meet the crop's N needs, manure and compost applied to match P-removal-rates and six different rates of inorganic fertilizer (0, 50, 100, 150, 200 and 250 lb of N/acre) applied at V5. All plots received 20 lb of N/acre at planting as starter fertilizer. The manure and compost plots were further split in two subplots (split plot design) with addition of 150 lb of sidedress N/acre to one of the two subplots.

In addition, a randomized complete block design was added to the trial area with five different rates of inorganic fertilizer (0, 50, 100, 150 and 200 lb of N/acre) applied pre-plant and replicated four times. The same design was used at three additional locations in central New York. In all plots a GreenSeeker 505 handheld proximal sensor coupled with a GPS-data logger (Nomad, Trimble) was used to map the canopy reflectance, measuring the NDVI, on a weekly basis starting at V3 until V12 growth stage (nine measurements total). Each plot was harvested for silage (2 rows of 40 feet per plot) and for grain. Preliminary results from three of the trial locations indicated that NDVI was a good predictor of both corn silage and grain yield and that the best timing to scan was between V7 and V8. Complete analyses will be presented, including yield and NDVI data from all sites.

ASSESSING SOIL BIOLOGICAL RESPONSE TO HARDWOOD BIOCHAR IN A TEMPERATE CLIMATE AGRICULTURAL SOIL. E.J. Cole*, M. Hashemi, B. Xing, and J. Blanchard, Westfield State University, Westfield, MA (249)

ABSTRACT

Soil quality has become a major factor used in assessing sustainable land management and the overall environmental quality, food security, and economic viability of agricultural lands. Recently, biochar has been touted as having many potential uses as a soil amendment for improving soil quality, specifically improving cation exchange capacity, pH and nutrient availability. However, soil biology also plays a significant role in biogeochemical processes that influence soil health and should be included in a more comprehensive study of soil health. A three-year field trial was conducted with the cumulative purpose of better understanding the effect that the application of a hardwood biochar has on soil health and quality. Assessment of soil biological, chemical and physical properties as affected by the addition of the hardwood biochar to agricultural soils showed significant changes in soil pH, cationic nutrient content as well as changes to the nematode community assemblages and bacterial community composition (identified through 16S rRNA sequencing) in the biochar-amended soils. While overall community diversity was not significantly affected, the abundance of specific bacterial taxa were significantly affected, indicating the potential for shifts in biogeochemical cycling in biochar-amended soils.

IMPACT OF MANURE INJECTION ON SOIL NITROGEN AND FATE OF MANURE ANTIBIOTICS. R. Maguire*, S. Kulesza, K. Xia, K. Knowlton, and P. Ray, Virginia Tech, Catawba, VA (250)

ABSTRACT

Manure from confined animal feeding operations has historically been surface applied, but this leads to losses of ammonia and risk of pollutant losses in surface runoff. Sometimes manure is tilled into soils, but this is not possible in no-till and forage systems. Therefore, there is increasing interest in low disturbance manure injection below the soil surface, to prevent ammonia volatilization, capture odors, and prevent losses of nutrient and other pollutants in surface runoff. Equipment is available for injection of liquid manure, such as dairy manure, and prototypes have been developed for dry manures such as poultry litter. This presentation will summarize several years of work on manure injection carried out in Virginia. Results will include reporting effects on nutrient and antibiotic losses in runoff, as well as ammonia losses and nitrogen cycling.

INVASIVE SPECIES

RHUS TYPHINA OUTCOMPETES *AILANTHUS ALTISSIMA* UNDER DIFFERENT LIGHT AND DENSITY CONDITIONS. C. Huebner*, J. Butnor, and R. Eaton, Northern Research Station, US Forest Service USDA, Morgantown, WV (252)

ABSTRACT

Ailanthus altissima is a prolific nonative invasive tree invading disturbed areas in the eastern United States, including managed forests. Its success is often attributed to a superior competitive ability. The competitive ability of *A. altissima* and an aggressive, early-successional native tree, *Rhus typhina*, was evaluated in a greenhouse and common garden. Tree height and shoot and root biomass of each species were measured under different density combinations and light levels in the greenhouse over 3 months, and the same densities were examined in the common garden over 2 growing seasons. In the greenhouse and the common garden, *R. typhina* grew faster and had greater shoot and root biomass than *A. altissima* in every density combination and light treatment. The root-to-shoot ratio of *A. altissima* was higher than that of *R. typhina* under greenhouse high-light treatments; preliminary data show the same in the common garden at density combinations with more *A. altissima* stems than *R. typhina*, suggesting *A. altissima*, given adequate light, puts more energy into root development than *R. typhina*. *Rhus typhina* and similar early successional natives may be able to prevent invasion by nonnative invasives like *A. altissima* at disturbed sites, highlighting the importance of maintaining native seed and seedling banks.

HEMLOCK WOOLY ADELGID: RESPONSE COORDINATION AND BIOCONTROLS IN NEW YORK STATE. C. Marschner* and M.C. Whitmore, Cornell University, Ithaca, NY (258)

ABSTRACT

Hemlock Woolly Adelgid, *Adelges tsugae*, is an invasive insect that has caused substantial mortality of Eastern hemlock (*Tsuga canadensis*) since its first detection on the US east coast in the early 1950's. It was first found in New York State in the late 1980's, where it has caused extensive mortality in the Catskills and is poised to enter the hemlock-dominated Tug Hill and Adirondack regions. Management practices involve the use of systemic insecticides to keep valuable hemlock resources alive while more long-term biological controls are established. The Hemlock Initiative is launching a collaborative effort to coordinate Hemlock Wooly Adelgid management in New York through prioritizing significant hemlock stands, coordinating volunteers for detection and to monitor tree health, developing a regional management plan, and developing biocontrol resources around the state. With aggressive coordination and effort, we hope to conserve New York's remaining hemlock resources.

INVASION ECOLOGY AND CLASSICAL BIOLOGICAL CONTROL OF THE EMERALD ASH BORER IN THE U.S.: RECENT PROGRESSES AND CHALLENGES. J.J. Duan* and D.E. Jennings, U.S. Department of Agriculture, Newark, DE (260)

ABSTRACT

The emerald ash borer (EAB), *Agrilus planipennis* Fairmaire, is a serious invasive beetle that was accidentally introduced to North America from northeast Asia probably in the mid 1990s. As of October 2015 EAB has spread to 25 states in the U.S. and two Canadian provinces, and has killed millions of North American ash trees (*Fraxinus* spp.) across the continent. Three biocontrol agents, one egg parasitoid (*Oobius agrili*) and two larval parasitoids (*Tetrastichus planipennisi* and *Spathius agrili*) originated from northeast China were approved in 2007 for field releases against *A. planipennis* in the U.S. Recently (June 2015), an additional larval parasitoid (*Spathius galinae*) collected from the Russian Far East was also approved for field releases against EAB in the U.S. Here we first review the progress made in the past decade with regards to the invasion ecology of *A. planipennis*, particularly the temporal and spatial dynamics of its spread and population dynamics in relation to various stages of its invasion process. We then present data on the population dynamics of EAB and associated natural enemies (native and introduced) in both the leading edge and aftermath forests, and demonstrate the potential and challenges in successful control of EAB in the U.S. via classical biological control programs.

MAJOR CLIMATIC NICHE SHIFTS IN OVER ONE THOUSAND INVASIVE SPECIES. D.Z. Atwater* and J.N. Barney, Virginia Tech, Blacksburg, VA (261)

ABSTRACT

Does a species retain its niche upon arrival onto a new continent? This question is central to our understanding of the ecological function of a species, and our ability to assess invasion risk of introduced species. How much and how often the niches of species change upon introduction remains the subject of considerable, active debate, and current evidence is inconclusive and often contradictory. In this study we use species distribution models to reconstruct changes in the realized climatic niches of over one thousand terrestrial plant species upon introduction to new continents. We use established as well as new techniques to characterize niche changes and to account for possible sources of sampling bias in over thirteen million species presence records across six continents. Our results suggest that the realized climatic niches of introduced species change dramatically upon introduction to a new range. For most species niche overlap between native and introduced populations fell below that expected among species (Schoener's D < 0.4), suggesting extreme differentiation. We find that the magnitude of climatic niche shifts depend on the life history and growth form of the introduced species, with annual and biennial species showing the smallest niche shifts, and woody species showing the largest shifts. This appears to be due to lags in the ability of longer-lived species to colonize novel habitats. We also find that niche shifts in introduced species reflect differences in climate availability between continents, as species were more likely to favor common climate in their introduced range. This tendency was strongest in shorter-lived species. Finally, even within their native range, species niches were not consistent across continents, although cross-continent niche shifts were smaller within a species native range than when comparing its native and introduced ranges. Overall we find evidence for major deviations in the native- and invasive-range climatic niches of an unprecedented number of terrestrial plant species, shedding new insight into how species respond to climate variability on new continents, and complicating our ability to assess the invasion risk of introduced species.

THE INVASIVE VASCULAR FLORA OF SIDEWALK PLOTS, BROOKLYN AND QUEENS, NY, 2007-2011. R. Stalter*, St. John's University, Queens, NY (262)

ABSTRACT

The vascular flora at 50 small sidewalk plots in Green Point Brooklyn, Kings County New York and St. John's University, Jamaica, Queens County, New York was identified and compared during Spring and Fall 2007 and through the Spring and Fall 2011 growing seasons. One hundred twenty one vascular plant species within 94 genera and 37 families have been identified. Eighty four taxa were identified in Brooklyn while 74 taxa were identified in Queens; many taxa at Brooklyn were also found at Queens as well. Ninety six taxa were identified at both sites in the spring while 55 taxa were identified in the fall. The Asteraceae (22 species) and Poaceae (18 species) were the largest families in the flora. Eighty five taxa, 69 percent of the flora are not native to the region. Thirty one taxa were observed only once during the five year study; those observed only once, twice and three times composed 50 percent of the flora. A parsimony algorithm was used to evaluate the distribution and co-occurrence of 121 vascular plant species identified at the above sites in Brooklyn and Queens in the fall and spring 2007 to 2011. One thousand plots were sampled during the five year study. The hierarchial cluster analysis forms two major clusters representing a Queens section, Queens Fall 2007 to Queens fall 2011, and a Brooklyn section from Brooklyn Fall 2007 to Brooklyn spring 2009. The data did not have enough resolution to produce a single unambiguous tree showing a unique hierarchial grouping of all sites. This may be explained by the dynamic nature of these “islands”, as many species colonize these islands briefly and disappear. The high diversity at the sidewalk plots may be a product of local disturbance.

BIOLOGICAL CONTROL OF MILE-A-MINUTE WEED (*POLYGONUM PERFOLIATUM*). D.R. Ellis* and C. Cheah, University of Connecticut, Storrs, CT (263)

ABSTRACT

Integrated Pest Management (IPM) approaches can be implemented for many species of invasive plants, including mile-a-minute weed (*Polygonum perfoliatum*), a highly invasive annual found in 13 eastern US states. Since it was first confirmed in Connecticut in 1997, mile-a-minute has spread to at least 42 municipalities in the state. It outcompetes and outgrows native species, causing ecological and economic harm. The vines can form dense mats, interfering with forest regeneration and seedling establishment. The biological control agent *Rhinoncomimus latipes* is a specialist on mile-a-minute as its host plant and has shown minimal impacts on non-target species following its release in the US. The integrated management of mile-a-minute weed is a collaborative effort between the University of Connecticut and the Connecticut Agricultural Experiment Station, and it involves multiple partnerships with local, state, regional, and federal stakeholders.

VEGETABLE AND HERB CROPS (I)

EFFICACY OF FUNGICIDES, BIOPESTICIDES, AND RESISTANT VARIETIES FOR MANAGING DOWNY MILDEW IN BASIL. M.T. McGrath*, Cornell University, Riverhead, NY (265)

ABSTRACT

Experiments were conducted from 2010 to 2015 under field conditions with naturally-occurring inoculum (wind-dispersed spores) to evaluate management practices for downy mildew, caused by *Peronospora belbahrii*, in sweet basil for conventional and organic production. Downy mildew has occurred every year on Long Island since 2008. Effective management was achieved with resistant experimental lines, in particular those bred by Pyne and Simon at Rutgers University and commercial lines tested in 2015. Conventional fungicides were effective applied following a preventive, 7-day schedule. Best control (at least 90 percent) was obtained with combination programs. Registered products used in these programs were Ridomil (soil drench after transplanting), Quadris, Ranman, Revus, and K-Phite, which was tank-mixed with Ranman and Revus. Biopesticides and other organic fungicides were not effective tested individually or in a program with multiple products applied on a once or twice weekly schedule. This may partly reflect challenge of delivering fungicide to the underside of leaves, where the pathogen develops. Control was not improved with an integrated program: fungicides applied to a partially-resistant variety. Plots were single rows of plants spaced 9 inches apart, Applications were made with a backpack sprayer using a boom with a nozzle directed down at the plant top plus drop nozzles on each side when plants were large. Downy mildew was assessed as percentage of leaves with symptoms, which is a rigorous assessment measure, but realistic reflecting zero tolerance for disease on fresh-market herbs.

SCREENING FOR GENETIC RESISTANCE TO FUSARIUM WILT FOR INTROGRESSION INTO SWEET BASIL. K. Homa*, Rutgers University, New Brunswick, NJ (266)

ABSTRACT

Sweet basil (*Ocimum basilicum*) is one of the most economically important herbs in the world that is favored for its flavor, fragrance, and culinary uses. Since the 2007 introduction of the basil downy mildew (BDM) pathogen, *Peronospora belbahrii*, 100 percent of the sweet basil acreage in the U.S. has been at risk. In addition, Fusarium wilt caused by *Fusarium oxysporum* f. sp. *basilicum* (FOB), which was first identified in the U.S. in 1990, is now resurging. Causing wilt, crown and root rot in sweet basil, once a plant becomes infected with FOB, death and crop loss follows. In response to the outbreak of this pathogen during the 1990s, several basil varieties 'tolerant' to FOB were introduced including 'Nufar' and 'Poppy Joe's'. To counter BDM and FOB, one of the most economically feasible strategies is to incorporate resistance to both pathogens into sweet basil. Inheritance of resistance to BDM was recently characterized in a full sibling family developed from a BDM resistant parent 'MRI' (P2) and BDM susceptible parent 'SB22' (P1). In 2014, FOB (10^6 conidia/mL) screenings revealed that MRI was highly susceptible to the virulent isolate FOB 33 (*Fusarium oxysporum* f. sp. *basilicum*; source Vannacci, Italy), while SB22 was substantially less susceptible, indicating that the MRI x SB22 family was also appropriate for studying genetic action for tolerance to FOB. In 2015, vegetative cuttings of the MRI x SB22 F2 generation were screened for tolerance to FOB 33. Individuals were rated for disease severity using a disease rating scale where 1 = resistant, 2 = mild leaf drop, 3 = stunted, wilted, 4 = side shoot regrowth and 5 = death. The frequency distribution among individuals of the F2 generation demonstrated a skew toward resistance suggesting dominant gene action conferred by SB22 for FOB tolerance. Chi-square tests were performed to determine goodness-of-fit to expected segregation ratios of multiple hypothesized major gene models in which 1 = resistant and 2-5 = susceptible. Evidence for goodness-of-fit was identified for the 3:1 ($\chi 2 = 3.11$; $P = 0.08$) and 13:3 ($\chi 2 = 0.111$; $P = 0.74$) gene models, suggesting tolerance from SB22 is controlled by one or two genes. In order to better elucidate gene action, additional generations of the MRI x SB22 family will need to be screened against FOB 33. These preliminary results can aid in the design of appropriate breeding and selection strategies for the development of new FOB resistance in sweet basil and incorporation of FOB resistance into the BDM resistant sweet basil both which are currently being developed at Rutgers University.

RESIDUAL ACTIVITY OF NEW INSECTICIDES FOR EUROPEAN CORN BORER CONTROL IN SNAP BEAN. R.A. Schmidt-Jeffris* and B.A. Nault, Cornell University, 14456, NY (267)

ABSTRACT

The processing snap bean industry views European corn borer (ECB), *Ostrinia nubilalis* (Hübner) larvae as contaminates and has zero tolerance for larvae in processed beans. Currently, the common practice of growers is to apply 1-2 pyrethroid treatments for ECB control. The objective of this study was to compare the ECB control efficacy of residues of newer, reduced risk materials to traditional pyrethroid treatments and an untreated control. The materials evaluated were Belt SC (flubendiamide), Besiege (chlorantraniliprole + lambda-cyhalothrin), Brigade 2EC (bifenthrin), Coragen (chlorantraniliprole), and Exirel (cyantraniliprole). ECB neonates were exposed to snap bean leaves with 1, 7, 10, and 14 d old field-aged residues for 3 d in laboratory bioassays and evaluated for mortality. The field where the snap beans had been treated was artificially infested with neonates to examine the field efficacy of each material × residue-age treatment. All products caused >80 percent mortality at 1-7 DAT. Coragen and Besiege both caused ~90 percent mortality even at 10 DAT. In the field, all material × residue-age combinations significantly reduced the percentage of plants and pods damaged compared to the check, except 14 DAT Belt SC. Most Coragen treatments and Besiege 1 DAT produced significantly higher marketable yield (lbs) than the check. These results indicate that newer materials with more selective modes of action (the diamides: Belt SC, Coragen, Exirel, and Besiege) are as effective as the industry standard (Brigade 2EC), even at unusually high pest pressure. Additionally, due to the duration of residue effects, only one application may be sufficient.

A PUSH-PULL METHOD FOR THE CONTROL OF MEXICAN BEAN BEETLE IN LIMA BEANS. A.W. Leslie*, K.A. Hamby, and C. Hooks, University of Maryland, College Park, MD (268)

ABSTRACT

Lima bean growers deal with a diversity of crop pests and diseases and must rely mainly on synthetic chemicals for control. Integrated pest management of the Mexican bean beetle (MBB) *Epilachna varivestis* (Coleoptera: Coccinellidae) is hindered in part because of wide variation in observed yield loss at different population sizes. We tested a push-pull strategy to reduce pest pressures from MBB on lima beans and minimize the need for chemical control. For this method, French marigold (*Tagetes patula*) was used as a repellent (push) plant that was inter-planted with the lima bean cash crop. This was paired with border rows planted with wax snap beans (*Phaseolus vulgaris*), which are favored by MBB over lima beans, and would act to lure the insects (pull) off of the cash crop and act as a trap crop. When compared with a lima bean monoculture, the push pull treatment had significantly higher numbers of MBB in the wax bean border rows, proving the efficacy of the trap crop. MBB numbers did not differ between internal rows, suggesting that there was not a strong repellency effect of the marigolds. Yields were significantly lower in the push pull treatments compared to monoculture, which was likely due to competition of the lima bean plants with the taller marigolds. Future trials of this push-pull strategy should focus on optimizing the benefit of the trap crop borders, while minimizing shading by the marigold plants.

INVESTIGATION OF THE PHYSIOLOGICAL EFFECTS OF HEAT STRESS ON LIMA BEAN (*PHASEOLUS LUNATUS*). E.G. Ernest*, University of Delaware, Georgetown, DE (269)

ABSTRACT

Heat stress reduces yields of May and early June-planted lima beans (*Phaseolus lunatus*) on the Delmarva Peninsula. High temperatures during flowering and pod set can result in later harvest, lower yield and split sets. Breeding heat tolerant lima bean cultivars adapted to the Mid-Atlantic region is one aspect of the University of Delaware lima bean breeding program. Using temperature controlled chambers in the greenhouse six lima bean genotypes were evaluated under hot (27 °C) and cool (18 °C) night conditions. The amount of pollen shed onto stigma and style and yield components were measured. Genotypes differed in their response to heat stress, but overall plants grown under high night temperature conditions shed less pollen onto the stigma and style. Heat stressed plants also produced fewer seeds, and fewer seeds per pod. Some genotypes produce fewer pods per plant under heat stress. Reduced pollen production and/or release accounts for some of the heat related yield loss observed in lima bean.

MANAGEMENT OF STEMPHYLIUM LEAF BLIGHT OF ONION WITH FUNGICIDES IN NEW YORK. C.A. Hoepting* and S.J. Pethybridge, Cornell University Cooperative Extension, Albion, NY (270)

ABSTRACT

Stemphylium leaf blight (SLB) caused by *Stemphylium vesicarium* has emerged as an aggressive leaf disease of onion in New York. In severe epidemics, SLB causes premature plant mortality, storability is reduced and incidence of bacterial bulb rot can increase by 50 percent. Results from 2013-2014 indicated that protectant fungicides, chlorothalonil (FRAC M5), mancozeb (M3) and iprodione (2) had no activity against SLB, and efficacy of pyrimethanil (Scala, FRAC 9) was variable.

In 2015, a small-plot replicated field trial was conducted in a commercial field of muck-grown yellow direct seeded pungent onions to evaluate the relative efficacy of currently registered and pipeline fungicides for control of SLB. It was arranged as a randomized complete block design with 18 treatments and 5 replicates. Each plot was 5 ft x 15 ft with four rows of onions spaced 15 inches apart; evaluations were taken from only the inside two rows. Treatments included 15 fungicides with active ingredients belonging to FRAC groups 3, 7, 9, 11, 12, 22, 29 and M3. The individual components of premix products were evaluated at the same rates as applied in the premix. An untreated control with no pesticide applications and an untreated with only onion thrips and downy mildew maintenance sprays were included. Highest label rates of fungicides were applied weekly for 5 weeks starting at the first detection of SLB lesions until the onions were 50 percent lodged from Jul-18 to Aug-13.

A final SLB score was determined within each plot from six disease and plant health assessments made from Aug-5 to Aug-30. A higher SLB score indicated more severe disease. All fungicide treatments had significantly lower SLB scores than the untreated with no pesticides (score 345), which was significantly lower than the untreated check with maintenance sprays (score 303). Maximum disease control was obtained with Luna Tranquility (FRAC 9 + 11; score 152), Merivon (7 + 11; score 182), Inspire Super (3 + 9; score 188) and Fontelis (7; score 193). Plots receiving fungicides belonging to FRAC 11 (Quadris (score 285) and Cabrio (score 273)) were not significantly different than the maintained untreated. Products with only FRAC 3 active ingredients (Inspire and Tilt) and only FRAC 7 active ingredient (Endura) provided significantly poorer disease control than Luna Tranquility, but not significantly different from Inspire Super or Fontelis, and among each other. The only product with just FRAC 9 active ingredient was Scala, which provided mediocre control (score 231). SLB control was significantly improved when Scala was premixed with fluopyram (7) within Luna Tranquility. Another premix product with FRAC 9 was Switch (9 + 12; score 244), which performed similarly to Scala. Disease control in plots receiving Omega (29; score 232) and Gavel (22 + M3; score 245) were not significantly different to those receiving Scala and Switch.

In conclusion, Luna Tranquility was the most efficacious fungicide for controlling SLB in onions and this was attributed to the presence of fluopyram (7). Generally, products belonging to FRAC groups 3 and 7 provided the best control of SLB. The lack of efficacy from strobilurin products (11) is concerning and the subject of further studies.

DICKEYA, A NEW POTATO PATHOGEN IN MAINE AND ELSEWHERE. S.B. Johnson*, UMaine, Presque Isle, ME (271)

ABSTRACT

Severe stand losses and blackleg symptoms were present Maine during the 2014 field season. Again, in 2015, severe stand losses and blackleg symptoms were present in a number of states including Maine. Blackleg symptoms have been associated with infection by *Pectobacterium* spp. Field symptoms and losses in 2014 and 2015 were far greater than had been previously encountered with *Pectobacterium* spp. infection. PCR confirmed *Dickeya* spp. as being present in number of these instances. Nonemergence is one symptom of *Dickeya* infection. Data from winter grow-out tests of potato seed lots showed an increase in nonemergence plants over the two years. It is probable *Dickeya* has been in the Maine seed system, albeit at low levels, for over 5 years. The pathogen may have been present and misidentified with symptoms attributed to unfavorable weather conditions. Warm conditions and excessive water favors the spread and development of *Dickeya*. One means of pathogen spread may be physical transfer within and among seed lots during harvest.

ATP SWAB SAMPLING IN FRESH PRODUCE PACKINGHOUSES. M.V. Melendez*, Rutgers Cooperative Extension, Trenton, NJ (273)

ABSTRACT

Sanitation of product contact surfaces in a packing house are an important part of reducing food safety risk. Most fresh produce growers are not sampling the product contact surfaces to evaluate the effectiveness of their sanitation step. Swab sampling for both ATP and generic *E. coli* by Rutgers Cooperative Extension On-Farm Food Safety Team members allowed for evaluation of the effectiveness of sanitation methods used at participant farms. Recommendations were then made to improve sanitation procedures. Best management practices for product contact surface sanitation methods were developed and used to educate fresh produce growers.

Five farms from New Jersey participated in the product contact surface sampling project. These farms varied greatly in their size, production methods, sanitation practices and commodities packed. Commodities included baby greens, onions, green beans, apples, and tomatoes. Packing line surfaces were swab sampled for both ATP and generic E. coli. Swabs were taken during active packing and after the line had been sanitized. Sanitation practices included the products Simple Green®, Sanidate® and a 10 percent bleach spray solution.

Sanitation practices in general reduced organic matter on the produce contact surface, reduced ATP levels and when generic *E. coli* was found reduced the number to 0. Variations in the sanitation step effectiveness were noted, particularly with the post sanitation step ATP numbers. Inconsistent use of the sanitation product is likely to blame. One farm found that its sanitation step was not effective in reducing ATP levels or eliminating generic *E.coli* numbers. It was determined that supervising staff were not present or were in a period of transition when ATP and generic *E. coli* levels on packing surfaces were not being managed properly.

Fresh produce growers are encouraged to create a standard operating procedure when developing their product contact surface area sanitation program. Training employees and posting directions in the work area are an important way of assuring proper sanitation methods are being used and that there is little variation in the effectiveness of the sanitation step. During times of transition it is important to consider relief staff tasks and ensure they are properly trained or supervised.

EPIDEMIOLOGY OF *POTATO VIRUS Y* IN NORTHERN MAINE.

A. Alyokhin* and A. Buzza, University of Maine, Orono, ME (274)

ABSTRACT

Potato virus Y (PVY) is an aphid-borne pathogen that causes significant crop losses throughout the U.S. It is non-persistently transmitted by at least 50 different aphid species, and can infect approximately 120 plant species in at least five different taxonomic families. Insecticides are not particularly effective in preventing its spread because no insecticide kills non-colonizing aphids quickly enough to prevent their interplant movement and probing. Twice during the growing season, samples were taken from plants other than potatoes growing in the vicinity of potato fields in northern Maine and comprising the majority of grassy broadleaf species in those areas. The samples were assayed for PVY using commercially available ELISA kits. Eight out of the 24 plant species surveyed were seropositive for PVY. Infection rates varied between 2 – 93 percent and were usually higher later in the season. Colonizing wingless aphids were counted weekly on fifty randomly selected plants within each field. Flight of colonizing and non-colonizing winged aphids was monitored using one green tile pan trap, one yellow pan trap, and two different sizes of yellow sticky cards deployed in random order within each field. Adult winged aphids were collected weekly, counted, and, whenever possible, identified to a species. Numbers of wingless colonizing aphids on sampled plants were generally very low. Sticky cards captured considerably more aphids compared to pan traps. Between pan traps, yellow ones were more productive than green tile ones. The black bean aphid, Aphis fabae, was the most abundant species, accounting for 51 percent of all captures.

AGRONOMIC CROPS (II)

WINTER COVER CROP STRATEGIES FOR MANAGEMENT OF HORSEWEED IN NO-TILL GRAIN SYSTEMS.
J.M. Wallace*, W.S. Curran, M. VanGessel, and D.A. Mortensen, Pennsylvania State University, State College, PA (277)

ABSTRACT

Glyphosate resistant horseweed (*Conyza canadensis*) is a significant weed management challenge for annual crop producers practicing conservation tillage in the Northeastern United States. In this region, horseweed typically behaves as a facultative winter annual, with distinct emergence periods in both fall and spring. Consequently, glyphosate-only burndown programs can become ineffective. Current management recommendations are to include multiple herbicide sites of action in burndown programs and to use soil residual herbicides if horseweed emergence is known to extend into the cash crop growing season. Looking forward, development of integrative approaches for horseweed management will be necessary to move selection pressure away from herbicides. No-till producers are increasingly integrating cover crops for soil health benefits, particularly following small grains or short season summer annual crops. This trend presents an opportunity to optimize cover cropping strategies for horseweed management. We conducted field experiments to evaluate cover crop strategies for horseweed management in 2014-2015 at Penn State's Russell E. Larson Agricultural Research Center (PSU-RELARC) in central PA and at University of Delaware's Carvel Research and Education Center (UD-CREC) near Georgetown DE. Cover cropping treatments were evaluated following small grain production and were imposed as a RCBD with a split-plot and four replications. Main plots were cover crop treatments: no cover, cereal rye (134 kg ha^{-1}), spring oats (134 kg ha^{-1}), cereal rye + hairy vetch (67 + 22 kg ha^{-1}), cereal rye + forage radish (67 + 6 kg ha^{-1}), spring oats + hairy vetch (67 + 22 kg ha^{-1}), and spring oats + forage radish (67 + 6 kg ha^{-1}). Split-plots were fertility treatments: 0 or 67 kg N ac^{-1} using AMS. Cover crops were planted using a no-till grain drill on 19-cm row spacing following burndown and fertilizer applications in early September. Cover crops were terminated at the cereal rye boot stage (Zadok 45) using glyphosate + 2,4-D (1.26 + 0.56 kg ha^{-1}) and soybean was planted across the study. Prior to planting cover crops, locally-collected horseweed seed was distributed in permanently marked microplots (0.50 m^2) at an average rate of 5,400 seeds m^{-2}. Cover crop biomass (kg ha^{-1}) and horseweed density and size were collected 10 weeks after planting (WAP) and at spring burndown. At PSU-RELARC, the addition of fertilizer (67 kg ha^{-1}) had a significant effect on cover crop biomass 10 WAP across cover crop treatments, which resulted in decreased horseweed density (216 plt m^{-2}) compared to cover crops without fertilizer (608 m^{-2}). Within fertilized plots, each cover crop treatment significantly decreased horseweed density 10 WAP by at least 45 percent compared to the control. Cereal rye + radish resulted in the greatest decrease (86 percent). Within unfertilized plots, cover crops did not decrease horseweed densities 10 WAP relative to the control, but did decrease the average diameter of horseweed rosettes. Similar trends were observed at spring burndown. In fertilized plots, cover crop treatments significantly decreased horseweed density relative to the control except for oats +

vetch. Horseweed density was notably lower in the rye and rye + radish treatments, 16 and 32 plt m^{-2} respectively, compared to the control (340 plt m^{-2}). Horseweed densities were low across the UD-CREC site 10 WAP, which precluded analysis of cover crop effects. At spring burndown, cover crop treatments influenced horseweed density similarly across fertilizer treatments. Only rye + vetch significantly decreased horseweed density (0 plt m^{-2}) compared to the control (80 plt m^{-2}), though the rye and oats + vetch treatments resulted in high levels of horseweed suppression (< 10 plt m^{-2}). Our results suggest that fall-planted cover crops have the potential to significantly suppress horseweed populations, thereby reducing herbicide selection pressure. General trends suggest that winter-hardy cover crops that produce high levels of ground cover 5 WAP provide greater horseweed suppression than winter-killed cover crops. Our results also indicate that N fertilization of cover crops may be necessary to maximize weed suppression benefits for horseweed management.

CHANGES IN SOIL QUALITY WITH INCREASING TIME IN NO-TILL PRODUCTION.
J. Jemison*, R. Kersbergen, and C. Majewski, University of Maine, Orono, ME (280)

ABSTRACT

No-till corn silage production is increasing in areas once thought to be too cold for such production methods. In the spring and summer of 2015, we interviewed early adopter no-till farmers about why they chose to convert, and sampled three to five fields per farm to assess whether length of time in no-till correlated with higher soil quality. Some of the farmers still lightly disked some of their fields, and so the soil quality of those fields were used for comparison. Length of time in no-till ranged from zero to five years. Using a standard bulb planter sampler, we took 12 cores per field to a depth of 15 cm. Earthworm midden counts, soil penetrometer depth to 300 psi, and cover crop use were included in the analysis. The primary reasons farmers noted for adopting no-till were labor, time and fuel savings. Chief concerns centered around weed management, particularly perennial weed management. Soil quality measures evaluated included active respiration (CO2 burst), labile amino nitrogen, water soluble carbon, and agrregate stability ranged greatly over the 20 growers and 75 fields sampled. Relationships of specific test parameters and length of time in no-till will be discussed in further detail. Its an interesting time of change for forage producers in New England.

ALFALFA-GRASS EVALUATION USING NIRS OR VISUAL ESTIMATION.
E. Karayilanli*, D. Cherney, P.K. Sirois, D.M. Kubinec, and J.H. Cherney, Suleyman Demirel University, Isparta, Turkey, NY (281)

ABSTRACT

Unlike most of the USA, NY and much of the Northeast rely on alfalfa grown in binary mixtures with perennial grasses, about 85 percent of the alfalfa in NY is sown with perennial grass. Both alfalfa and grass cultivars have been recently released that are potentially much higher in forage quality than normal types, but optimum time of harvest still depends on the ratio of alfalfa and grass in the mixture. Past studies have show that near infrared reflectance spectroscopy (NIRS) instruments can be effectively calibrated using known legume percentages of mixtures, but validation of these calibrations using data from other years or sites typically fails. Our objective was to evaluate whether NIRS and visual photo evaluation can effectively estimate the alfalfa:grass species ratio in mixed stands. Fresh and fermented samples were collected in 2012, 2013 and 2015, and samples for calibration were mixed to range from 0 to 100 percent alfalfa, on a dry weight basis (N = 1360). NIRS analysis and calibration was completed by Dairy One Cooperative, Inc. In spring and early summer of 2015 we acquired samples (n = 207) of alfalfa-grass stands in farmer's fields, and determined alfalfa and grass dry matter proportions for each sample after capturing a digital image (5-Megapixels) of the sampling area. A set of calibration photographs was identified that covered the range of alfalfa percentage in hand-spearated samples, selecting photographs that visually represented a decreasing alfalfa percentage, and also agreed with hand separation results. There was no benefit in separate fresh and fermented sample calibrations, a single calibration had an R^2 value of 0.994. Validation with a separate set of samples was successful with an R^2 value of 0.991. Calibrated visual estimates ($y = 13.3 + 0.833x$; $R^2 = 0.70$) tended to overestimate alfalfa when the alfalfa percentage of the stand was low.

ESTIMATING UNDIGESTED NDF AND RATE OF DIGESTION IN FORAGES. M. Valentine*, E. Karayilanli, J.H. Cherney, and D. Cherney, Cornell University, Ithaca, NY
(282)

ABSTRACT

Long-term in vitro digestions of forages, regardless of the procedure, provide estimates of rate of digestion of neutral detergent fiber (NDF) and also undigested NDF (uNDF) at some time point shorter than infinity. The indigestible fraction needs to be accurately estimated, as it must be removed in order to determine rate of digestion of the potentially digestible fraction. Potentially digestible particles can be retained, or indigestible particles can be lost, depending on the filtration procedure. The two most common methods for in vitro digestion are the conventional flask procedure and the ANKOM filtration bag procedure. The possibility exists in long-term digestions for small indigestible particles to escape the ANKOM F57 filter bag, due to its pore size of 25 μm. A filter bag has been developed (F58) with a pore size of 8-10 μm, decreasing the chance of losing small indigestible particles during the NDF procedure, but has not yet been tested for in vitro digestions. Twenty four forage samples representing a wide range of temperate and tropical grasses and legumes were collected. Duplicate samples in both F57 and F58 filter bags were digested for 30, 120 and 240 h, and the entire digestion procedure was repeated a second time. Cumberland Valley Analytical Services digested the same sample set using the conventional flask procedure three times on separate days. Preliminary in vitro runs with six time periods showed that most forages did not differ in uNDF between 96 and 240 h of digestion, except for corn silage and warm-season perennial grasses. uNDF values ranged from less than 100 to greater than 500 g/kg, depending on the forage source. Both uNDF and rate of digestion were similar using F57 and F58 filter bags. Rates of digestion varied greatly among forages, but were relatively consistent over procedures.

WEED CONTROL AND CROP SAFETY WITH METRIBUZIN IN WINTER WHEAT. M. VanGessel*, Q. Johnson, and B. Scott, University of Delaware, Georgetown, DE (283)

ABSTRACT

Wheat and barley farmers in the Mid-Atlantic region have relied extensively on acetolactate synthase-inhibiting herbicides for both broadleaf and grass weed control. Over-use of these herbicides have resulted in herbicide-resistant Italian ryegrass and common chickweed, as well as an increase in species less susceptible to these herbicides. UD Weed Science Program has evaluated other herbicide mechanism of actions to identify herbicides to assist with managing these problematic species. Metribuzin has received a 24c label for use in small grains in many of the Mid-Atlantic States for management of ALS-resistant common chickweed. In addition, metribuzin provides excellent control of henbit, ivyleaf speedwell, and cornflower. However, there has been concern about wheat injury when metribuzin is used. Metribuzin is labeled for application as early as spike stage in pre-packaged mixture with flufenacet or as soon as 2-leaf stage when applied by itself. The objective of this experiment was to determine how application timing of metribuzin might influence winter wheat safety. The study was conducted at the UD Research and Education Center near Georgetown for three years on loamy sand soils with organic matter content of 0.8 to 1.5 percent and soil pH from 5.9 to 6.1. Winter wheat ('Shirley' or 'USG3555') was drilled on October 23, 2012, October 28, 2013, or October 20, 2014. Treatments were a factorial design of metribuzin rate (0.094 or 0.187 lb ai/A [2 or 4 oz of product/A]) and timing based on wheat development (preemergence, 2-leaf stage, early spring at "green-up", or spring [3 weeks after the early spring timing]). An untreated check was also included. Treatments were arranged as a randomized complete block design with three replications. All metribuzin treatments included a non-ionic surfactant at 0.25 percent v/v. The sites were treated with thifensulfuron:tribenuron to reduce weed competition and sites for 2013/2014 and 2014/2015 seasons were supplemented with irrigation to prevent moisture stress. Visual ratings of injury (leaf burn or stunting) were determined throughout the growing season, and plots were mechanically combined to determine yields at wheat maturity. Yields were converted to percent of untreated check for analysis.

Fall ratings were significant only in 2013. The main factors of rate and timings were significant, with the higher rate or 2-lf application timing causing more injury. Ratings taken the third week of April each year had a rate by timing interaction. Injury for 2012/2013 season was less than 10 percent, with high rate of metribuzin applied at the spring timing exhibiting the most stunting. In the 2013/2014, high rate of metribuzin applied at 2-lf stage had 63 percent stunting and low rate at 2-lf stage was 25 percent. High rate of metribuzin applied PRE was 18 percent while low rate was 9 percent. All other treatments had less than 7 percent stunting. In 2014/2015 season high metribuzin rate at PRE or 2-lf stage was over 53 percent stunting, low metribuzin rate applied at 2-lf stage was 37 percent and low rate applied PRE was 16 percent. All other treatments were less than 2 percent stunting. Yield, as percent of untreated check, was highest with early spring treatment all years, PRE timing in 2012/2013 and 2013/2014, 2-lf timing in 2012/2013 only, and spring timing in 2013/2015 only.

Metribuzin can improve control of herbicide-resistant and troublesome species in small grains. Applications in early spring were the safest timing to minimize stunting and consistently maintain winter wheat yields.

SUB-LETHAL DICAMBA DOSE IMPACT ON GROUP V SOYBEAN GROWTH AND YIELD. A.M. Growe*, M.K. Bansal, J. Copeland, J.T. Sanders, B.W. Schrage, L. Vincent, and W.J. Everman, North Carolina State University, Raleigh, NC (287)

ABSTRACT

With the spread of glyphosate resistant weed species throughout North Carolina, there has been a renewed interest of using auxin herbicides for weed control options in the state. As dicamba, a common auxin herbicide, is being incorporated back into herbicide programs, there is concern of off target movement to sensitive crops in adjacent fields. Although dicamba drift has been widely researched, to date, there has been little information reported on soybean varietal responses to dicamba at sub-lethal rates.

The objective of this study was to evaluate the effects of sub-lethal rates of dicamba on various group V soybean cultivars at vegetative and reproductive growth stages. Effects of dicamba were determined by collecting visual injury ratings, height reductions and yield. Experiments were conducted in Upper Coastal Research Station (Rocky Mount, NC) and Caswell Research Station (Kinston, NC) during 2015. Treatments were arranged in a five by seven by two factorial randomized complete block design with five soybean varieties treated with dicamba at 1.1, 2.2, 4.4, 8.8, 17.5, 35, and 70 g ae ha^{-1} (1/512 to 1/8 of the labeled use rate for weed control in corn) during V4 and R2 growth stages. All data were subjected to analysis of variance and means were separated using Fisher's Protected LSD at p= 0.05.

A wide range of visual injury was recorded at 1, 2 and 4 WAT for all 5 varieties and both timings. Increasing levels of injury were associated with increasing dicamba rates for all varieties. The V4 injury ratings ranged from 17-69 percent 1 WAT. Significant differences in height reductions, when compared to the non-treated, check were also observed. Height reductions were more severe at the V4 timing than R2 for all varieties. Height reduction 4 WAT, for all varieties, ranged from 35-39 percent at the V4 timing and 13-26 percent for the R2 timing. Dicamba effects on soybean yield have not yet been determined. Conclusions from this study reveal the importance of making responsible dicamba applications so that risk of drift and volatility is minimized.

PALMER AMARANTH CONTROL AND SOYBEAN TOLERANCE TO BALANCE BEAN HERBICIDE. B.W. Schrage* and W.J. Everman, North Carolina State University, Raleigh, NC (288)

ABSTRACT

The integration of new technologies and management strategies is becoming increasingly necessary to control Palmer amaranth in Southeastern soybean production. The anticipated release of the isoxaflutole-based product Balance Bean by Bayer CropScience and the anticipated deregulation of HPPD-tolerant soybean varieties, pending regulatory approvals, could serve as a new-era rotational tool enabling applications of isoxaflutole, glufosinate, and glyphosate. Best management practices dictate that rotating herbicide mechanisms of action can reduce weed seed bank densities—prompting further investigation of stacked traits.

In 2015, experiments were conducted in Clayton, South Mills and Sunbury, North Carolina to evaluate the impact of various herbicide programs including Balance Bean on Palmer amaranth and HPPD-tolerant soybeans. In Clayton, all herbicide treatments exceeded 90 percent control of Palmer amaranth for the entire growing season. In Sunbury and South Mills, eight application rates, ranging from 20 to 160 g ai ha^{-1} were applied to non-crop plots. Control of Palmer amaranth exceeded 90 percent in both locations at higher rates. Results suggest that pending the introduction of HPPD-tolerant soybean varieties bred for Southeastern soybean producers, Balance® Bean could become an influential aspect of herbicide programs that embrace rotational technology.

ENLIST™ CORN WEED CONTROL PROGRAMS IN THE MIDWEST. D. Simpson and O. Castello*, DAS, Lancaster, PA (289)

ABSTRACT

The Enlist™ weed control system is being developed in multiple crops including Enlist™ corn. Enlist corn has been extensively evaluated in field research trials since 2006 and was deregulated by the United States Department of Agriculture in September 2014. Enlist corn, stacked with SmartStax® technology, provides tolerance to both 2,4-D and glyphosate as well as above- and below-ground insect resistance. Enlist Duo™ herbicide with Colex-D™ technology is a proprietary blend of 2,4-D choline and glyphosate dimethylamine (DMA) developed by Dow AgroSciences for use on Enlist crops. Enlist Duo was registered with the U.S. Environmental Protection Agency in October 2014. Dow AgroSciences will be recommending the use of soil residual herbicides as a part of the Enlist system to provide early season weed control and crop yield protection along with additional modes of action to manage weed resistance. Field research trials were conducted in 2013 (20 trials), 2014 (12 trials) and 2015 (11 trials) to evaluate herbicide programs including Enlist Duo (2,4-D choline + glyphosate DMA) and SureStart® or SureStart II herbicide (acetochlor + clopyralid + flumetsulam) for weed control and crop tolerance. Treatments included SureStart applied preemergence (PRE) followed by a postemergence (POST) application of Enlist Duo to V4 corn, Enlist Duo + SureStart applied early postemergence (EPOST) to V2 corn, Enlist Duo +

SureStart applied POST to V4 corn, and Enlist Duo, by itself, EPOST and POST. SureStart PRE rate varied by soil type (1170 to 1750 g ae/ha) and SureStart EPOST and POST rate by protocol (875 to 1170 g ae/ha). Enlist Duo was applied POST (1640 and 2185 g ae/ha) following PRE applications of SureStart, as a tank-mix with SureStart applied EPOST and POST, or applied by itself, EPOST and POST At 28 days after the POST application timing, SureStart PRE followed by Enlist Duo POST provided greater than 95 percent control of glyphosate-resistant waterhemp, common ragweed, and giant ragweed and 98 percent or greater control of glyphosate-susceptible weed species. Enlist Duo + SureStart POST provided 97 percent or greater control of glyphosate- resistant waterhemp and common ragweed and 87 to 93 percent giant ragweed. POST Enlist Duo + SureStart treatments provided 94 percent or greater and 97 percent or greater control of glyphosate-susceptible weed species, respectively. Corn tolerance was evaluated 7 and 14 days after the POST applications. SureStart applied PRE followed by Enlist Duo POST averaged less than 2 percent visual injury 14 days after POST application. The tank-mix of Enlist Duo + SureStart POST resulted in 2 percent or less visual injury 14 days after POST application. Residual herbicides are an effective tool to prevent yield loss caused by early season weed competition and bring additional modes of action to the weed control program as a component of weed resistance management best practices. These trials demonstrate the utility of residual PRE herbicides followed by POST applications of Enlist Duo as part of the Enlist system in Enlist corn.

™® Enlist, Enlist Duoand SureStart, are trademarks of The Dow Chemical Company ("Dow") or an affiliated company of Dow. Enlist Duo herbicide is not registered for sale or use in all states. Contact your state pesticide regulatory agency to determine if a product is registered for sale or use in your state. Always read and follow label directions. SmartStax® multi-event technology developed by Monsanto and Dow AgroSciences, LLC. SmartStax® and the SmartStax logo are registered trademarks of Monsanto Technology, LLC.

AUTHORITY MTZ ALTERNATIVE BURNDOWN TANKMIXES IN SOYBEANS. J. Reed*, R. Ritter, M.L. Flessner, and M. VanGessel, FMC, Philadelphia, PA (290)

ABSTRACT

Authority MTZ DF is a 45 percent dry flowable formulation containing 0.18 lbs sulfentazone and 0.27 lbs metribuzin. Authority MTZ DF is labeled for use as a fall application and spring pre-plant and pre-emergence application ahead of corn and soybeans. The "burndown" herbicide application in reduced tillage, no-till and stale seedbed systems is critical to achieving and maintaining weed control to maximum yields. Over reliance upon glyphosate and ALS inhibiting herbicides has resulted in various weeds species becoming herbicide resistant to one or both modes of action. Recently, in the Southern delta states, postemergence applications of PPO postemergence herbicides has resulted in herbicide resistance in specific weed species. Use of effective modes of action in tankmixes with PPO herbicides, such as Authority MTZ DF in burndown tankmixes, has proven to be effective in safely overcoming herbicide resistant weeds.

TURF AND ORNAMENTALS

ANNUAL BLUEGRASS SEEDHEAD SUPPRESSION. J.A. Borger* and T.L. Harpster, The Pennsylvania State University, University Park, PA (291)

ABSTRACT

Studies were conducted on a mature sward of annual bluegrass (*Poa annua*) at the Valentine Turfgrass Research Center, Penn State University, University Park, PA. The objective of the studies was to determine if selected materials applied in the spring could suppress annual bluegrass seedhead populations under simulated golf course fairway conditions. The studies were randomized complete block designs with three replications each. Treatments were applied on 12 April (PRE-BOOT), 17 April (BOOT) 29 April (2 WAT) and 8 May (2WAA) 2015, using a three foot CO_2 powered boom sprayer calibrated to deliver 40 gpa using one, flat fan, TP9504EVS nozzle at 50 psi . The initial treatment was applied at the pre-boot stage of growth of the annual bluegrass. The test site consisted of approximately 95 percent annual bluegrass and 5 percent mix of creeping bentgrass (*Agrostis stolonifera*) and perennial ryegrass (*Lolium perenne*) at the initiation of the study. Annual bluegrass populations were visually evaluated for the percent seedhead coverage in order to evaluate the test material's ability to suppress annual bluegrass seedheads. The test site was mowed at 0.50 inches three times a week with a reel mower. Turfgrass was irrigated on an as needed basis to prevent moisture stress. Data was analyzed with ARM 8.5.0 using Duncan's New MRT at the 0.5 percent significant level. This research compared some different techniques and product combinations to enhance the annual bluegrass seedhead suppression. When Primo and Proxy are applied as a combination twice during the spring there have been varying degrees of efficacy over the years. In this research when Civitas One, Maintain CF125, and MCPP were in combination with Primo and Proxy combinations, generally annual bluegrass seedhead suppression was improved. More research is needed to better understand the biological mechanisms at work and exact product rates and application timings.

USING HERBICIDES TO RENOVATE TURF DOMINATED BY ANNUAL BLUEGRASS. B.S. Park*, C. Mansue, and J.A. Murphy, Rutgers, The State University of New Jersey, New Brunswick, NJ (292)

ABSTRACT

Annual bluegrass (*Poa annua* L.) is a highly opportunistic weed that frequently invades golf and sports turfs. Among cool-season turfgrasses, perennial ryegrass (*Lolium perenne* L.) can be effectively overseeded into existing turfs given its rapid germination and aggressive establishment. The objective of this study was to assess the effectiveness of herbicide programs designed to suppress annual bluegrass in turf overseeded with perennial ryegrass. A study was initiated on a loam during September 2015 on a mature stand of annual bluegrass in North Brunswick, NJ. The test area was core cultivated with soil reincorporated on 3 September 2015 . Treatments consisted of 13 herbicide programs and an untreated check arranged in a randomized complete block design with four replications. Programs involved applications of glyphosate, amicarbazone, mesotrione, and ethofumesate applied during 4 Sep. through 3 Dec. 2015 using a CO_2 backpack sprayer calibrated to deliver 375 L ha^{-1}. 'Manhattan 5 GLR' perennial ryegrass was slit-seeded at 527 kg ha^{-1} on 14 Sep. 2015. Perennial ryegrass establishment (0 to 100 percent scale; 100 percent=complete perennial ryegrass turf cover) and herbicide phytotoxicity (1 to 9 scale; 9=no phytotoxicity) were visually assessed during autumn 2015. Treatments resulting in the greatest perennial ryegrass populations (86 and 79 percent) on 9 Nov. 2015 were glyphosate (0.6 kg ai ha^{-1}; 4 September) followed by (fb) ethofumesate (2.3 kg ai ha^{-1}; 6 October and 3 November) and glyphosate (0.6 kg ha^{-1}; 4 September), respectively; neither treatment caused phytotoxicity greater than the check though 9 Nov. 2015. Treatments that caused the most severe phytotoxicity on 9 Nov. 2015 were mesotrione (0.14 kg ai ha^{-1}; 4 September) fb mesotrione (0.1 kg ha^{-1}; 15, 19, 22 and 26 October); mesotrione (0.14 kg ai ha^{-1}; 15 October) fb mesotrione (0.1 kg ha^{-1}; 19, 22, 26, and 30 October); and glyphosate (0.6 kg ha^{-1}; 4 September) fb mesotrione (0.14 kg ai ha^{-1}; 15 October) fb mesotrione (0.1 kg ha^{-1}; 19, 22, 26, and 30 October). Trial evaluation will continue through spring 2016; however, initial results suggest that nonselective control of annual bluegrass before overseeding was the most effective herbicide strategy to increase perennial ryegrass during the first months of establishment.

CHALLENGES ASSOCIATED WITH MEASURING THE INFLUENCE OF GREENS CANOPY ANOMALIES ON BALL ROLL CONSISTENCY.

S.D. Askew*, S.S. Rana, and J.R. Brewer, Virginia Tech, Blacksburg, VA (293)

ABSTRACT

Although the sole intent of putting a golf ball across a green is directionally driven, no peer-reviewed research has investigated surface factors that influence ball direction on greens. Greens surface anomalies, like annual bluegrass (*Poa annua*), have long been reputed to impact ball roll direction and many golfers blame these factors for missed putts. So why the lack of scientific evidence? After two years of research and rolling over 10,000 golf balls in the laboratory and on several Virginia golf courses, we have come to realize that several sources of error must be minimized in order to detect the subtle influence that surface anomalies have on ball roll direction. We have found these sources of error to severely limit the utility of commercially-available ball roll devices for the purpose of measuring the consistency of simulated golf putts. The first is ball center of gravity. Although off-center golf balls can roll erratically, we have found that 180 golf balls from 13 manufacturers representing popular brands and a range of dimple patterns were all near perfect center of gravity as assessed using the brine-solution method. The second error source is terminal deceleration. Golfers are typically trained to putt balls such that a missed putt will leave the ball 30 to 45 cm past the cup to avoid erratic ball movement that occurs when the ball's momentum can no longer overcome frictional forces. For ball dispersion measurements in the field, we overcame this problem by placing pressure-sensitive paper on a strike plate positioned 30 cm short of the total putt distance. The third source of error was ball roll legacy or "tracking" effects. On two creeping bentgrass greens mown at 3 mm, balls repeatedly rolled in the same direction tended to roll further in a curvilinear trend, reaching 20 to 30 cm more roll distance when preceded by just 3 balls. In addition, balls rolled repeatedly in the same direction tended to stay in the "track" formed by previous balls and had statistically less directional variation than when the canopy was brushed between each ball roll. The final error source was erratic ball direction during simulated putts. We selected three commercial ball roll devices [USGA Stimpmeter (USM); Greenstester (GT); and Pelz Meter (PM)] and three custom-made devices [Putt Robot (PR); Carpet Ramp (CR); and Flexible Ramp (FR)] to test ball roll consistency of 13 golf ball models from leading manufacturers. Dispersion was measured by rolling the balls at stimp distance (320 cm) over a synthetic carpet designed for outdoor putting greens, photographing the resting position of each ball, and determining the pixel coordinates of the ball center using SigmaScan software. The PR and FR were the most consistent, having a directional skew resulting in a 16 and 17 mm deviation from the median ball resting position. The deviation caused by GT, USM, and PM was 32, 41, and 42 mm, respectively. Assessments with a high-speed camera suggest that ball oscillation is the likely cause of erratic ball direction from GT, USM, and PR. In field studies, we were able to use the PR while correcting for potential error sources and show that small patches of annual bluegrass will statistically decrease consistency of ball roll direction. We could not detect any influence of annual bluegrass on ball directional consistency when using the GT, presumably due to inherent error caused by the device.

COMPARISON OF VEGETATION INDICES FOR CHLOROPHYLL CONTENT AND DROUGHT STRESS.

D.S. McCall* and D. Sullivan, Virginia Tech, Blacksburg, VA (294)

ABSTRACT

Yield is not an important characteristic for maintained turfgrasses, therefore stands are measured by aesthetic appearance and ability to withstand a variety of stresses. Historically, researchers have quantified the acceptance of a turfgrass by its visual quality. The National Turfgrass Evaluation Program recommends the use of a 1-9 subjective rating scale by a trained evaluator, where 9 equals highest turf quality and a minimally accepted threshold is defined by the rater. Due to rater inconsistencies and bias, vegetation indices (VI) that utilize light reflectance from plants at specific wavelengths have been adopted as an objective alternative measurement because of strong correlation to visual ratings. The normalized difference vegetation index (NDVI) is commonly used in agricultural research and related fields because of a strong relationship to many plant health characteristics. Relationships have been established between NDVI of turfgrass canopies and important factors that impact stand health, including soil moisture content and leaf chlorophyll content. Numerous VI have been established because of demonstrated relationships to specific variables of interest. It is not known whether indices are useful for distinguishing stress to turfgrasses by limited moisture or chlorophyll availability. Therefore the objective of this ongoing study was to compare the relationships of twelve established VI to tissue chlorophyll content and soil water content of creeping bentgrass. Eight week old 'L-93' creeping bentgrass

Conetainers were maintained under greenhouse conditions with a soil profile built to USGA specifications. A 2x2 factorial design was arranged in randomized complete blocks with five replications. The experiment will be repeated once. Factors included A) irrigated (1 cm $48hr^{-1}$) versus non-irrigated and B) fertilized (24.4 kg N ha^{-1}, 46-0-0) versus non-fertilized. Data collected included spectral reflectance (400-1100 nm), soil volumetric water content, leaf chlorophyll content, and visual assessment of turf quality and turf color. Data were subjected to analysis of variance and means were separated (LSD, P = 0.05). Relationship between variables was determined using Pearson's correlation coefficient. All VI were moderately to strongly related to visual turf quality (49-85 percent). NDVI and three other indices were useful for detecting treatment effects of both irrigation and fertilization, but were unable to differentiate. Four indices (WBI, GRVI, RGVI, NDRE) detected changes in soil moisture content but not fertility, whereas fertility effects could be distinguished with four different VI (CHLOR1, GNDVI, SUM G, NDVI 705). The strongest relationship to soil moisture was with the water band index (WBI, 80 percent). All distinguishable VI for fertility had the strongest correlation to turf color (78-85 percent) and chlorophyll content (55-63 percent). This research demonstrates that spectral data may be used to differentiate soil water availability from plant nutritional needs.

TAKING THE STING OUT OF TURF PROTECTION: A SOIL MOBILE ABAMECTIN FORMULATION FOR NEMATODE CONTROL. K. Greig*, D. Norton, C. Irwin, and M. Coulter, Vive Crop Protection, Toronto, ON (295)

ABSTRACT

Penetration of nematicides through turf, thatch and soil presents a unique problem for control of nematodes in turfgrass. Many of the products currently used are found to have minimal mobility in the soil due to their affinity for binding to organic matter and the lack of options for soil incorporation. In this study, an abamectin formulation with enhanced soil/thatch mobility was developed and evaluated. To demonstrate improved mobility and control of nematodes, laboratory studies were conducted in 2014 and 2015, followed by turf trials in summer 2015. To test mobility in the laboratory, soil columns were packed with either sandy or silty loam. The soil was saturated and formulations were added to the top of each column, followed by an artificial rainfall event. After drying, segments were analyzed using HPLC to determine movement of the active through the column. Laboratory trials suggested much more targeted movement of abamectin from experimental products than from commercial controls. Experimental formulations were found to penetrate much deeper into the columns, with no penetration beyond 0.25 m. Formulations were also applied to golf course greens in comparison to two commercial controls. Population counts of nematodes were assessed as well as root length and percent turf greening. Turf trials on golf course greens demonstrated better control of nematode populations and turf quality as compared to commercial standards. This experimental formulation has shown to be much more effective for control of nematodes in turf than the commercial products tested.

MANAGING GROUND IVY AND WILD VIOLET IN TURFGRASS WITH FLUMIOXAZIN.
J. Fausey*, Nufarm Americas, Fremont, OH (296)

ABSTRACT

Ground ivy (*Glechoma hederacea*) and wild violet (*Viola* spp.) are two common and consistently problematic perennial broadleaf weeds found in cool season turfgrass. Managing these weeds is a dilemma many lawn care operators, sod producers and golf course superintendents face. Ground ivy is a prolific aromatic weed in the mint (Lamiaceae*)* family. This weed has numerous medicinal uses, and is commonly used as a salad green in many countries. Collectively, most managers of turfgrass refer to common blue violet (*Viola sororia*), wooly blue violet (*Viola papilionacea*), and confederate violet (*Viola sororia f. priceana*) all as wild violet. Additionally, yellow violet (*Viola pubescens*) can be found infesting cool season turf. All of these violet species are perennials with heart shaped leaves and, once they are established, can be very difficult to control. Ground ivy and violet spread by rhizomes and by seed making them difficult to control with a single application of any herbicide. Over the past several years numerous active ingredients for managing these weeds have been evaluated, with some of these materials showing good activity, but they did not provide the needed level of tolerance to be used around ornamentals. The lack of a highly effective pre or postemergence herbicide continues to leave those in the turf industry with few means of complete control once they are established. One new option, Flumioxazin, has been evaluated in Kentucky bluegrass and perennial ryegrass and consistently displayed improved speed and long-term control of these weeds. Nufarm Americas is developing Flumioxazin for use in cool season turfgrass. Field experiments were conducted in 2015 evaluating the potential for using Flumioxazin herbicide to manage these weeds. The objective of these trials was to evaluate the performance of Flumioxazin herbicide when applied in combination with other active ingredients and under different environmental conditions to determine the potential for this herbicide in the cool season turfgrass market. In addition to evaluating Flumioxazin herbicide at several locations, treatments included evaluation of different rates, timings and combinations. Data from these trials confirmed Flumioxazin provides an alternative management strategy in control broadleaf weeds for lawn care operators, sod producers and golf course superintendents.

POST-EMERGENT CRABGRASS CONTROL IN LAWNS USING ALTERNATIVES TO SYNTHETIC HERBICIDES.
D. Linde*, S. DeBroux, and S. McDonald, Delaware Valley University, Doylestown, PA (297)

ABSTRACT

In recent years there has been some research published on using alternatives to synthetic herbicides to control broadleaf weeds post-emergently in turfgrass but little has been published or presented on using alternative herbicides to control crabgrass post-emergently. Therefore, a 2-year study was conducted to evaluate post-emergent crabgrass control in lawns using alternatives to synthetic herbicides. In 2014, 12 treatments arranged in a RCBD with 4 replications were applied to a lawn area in

Doylestown, PA that contained smooth crabgrass (*Digitaria ischaemum*) and a mix of cool-season turfgrasses. The study was repeated in 2015 on a different site in Doylestown. Crabgrass age at time of initial treatment ranged from the 4-leaf to 1-tiller stage. Treatments included hand-pull, fenoxaprop (synthetic herbicide), pelargonic acid, a citric acid & clove oil mix, FeHEDTA, citrus oil, a eugenol & phenethyl propionate mix, rosemary oil, soybean oil, vinegar, glyphosate then seed perennial ryegrass, and an untreated control. Treatments were applied to 1 m^2 plots as spot treatments for crabgrass using a hand-pump spray bottle and were re-applied 11 days after initial application to any remaining crabgrass. Glyphosate-treated plots were seeded with perennial ryegrass at 11 DAIT instead of reapplying glyphosate.

Percent crabgrass cover was evaluated approximately every 7 days after initial treatment (DAIT) for 49 days. Transformations did not stabilize variance for percent crabgrass cover data thus non-transformed data were subjected to ANOVA with means separated by Tukey's HSD. Untreated control plots were included in analysis. Turf quality was assessed visually according to NTEP standard practices where 9 was outstanding or ideal turf, 6 acceptable, and 1 was the poorest or dead turf.

Hand-pull, fenoxaprop, and glypshosate/seed were the only treatments in both years to significantly reduce crabgrass cover compared to the untreated control by 49 DAIT. FeHEDTA had significantly less cover than the control in 2015 but not in 2014 by 49 DAIT. Various treatments (pelargonic acid, vinegar, citrus oil, eugenol & phenethyl propionate mix, rosemary oil, soybean oil, citric acid & clove oil mix) reduced crabgrass during the first few days after application but then the crabgrass recovered to levels equal to the control. These products injured the turfgrass as well as the crabgrass. Also, most treatments caused thinning of the turf and crabgrass canopy which led to additional crabgrass germination. Hand-pull and fenoxaprop were the only treatments that significantly reduced crabgrass without a major decline in turf quality. Therefore, hand-pull was the only alternative to synthetic herbicide treatment that provided acceptable crabgrass control.

RESISTANCE MECHANISMS TO ACCASE INHIBITORS IN GOOSEGRASS FROM GEORGIA.

P. McCullough*, J. Yu, and P. Raymer, University of Georgia, Griffin, GA
(298)

ABSTRACT

A goosegrass biotype with suspected resistance to ACCase-inhibitors was identified in Georgia. The objectives of this research were to evaluate (1) the resistance level of this biotype to ACCase-inhibitors, (2) efficacy of various herbicide for control, and (3) the physiological and molecular basis of resistance. In greenhouse experiments, the rate of diclofop-methyl that reduced dry shoot biomass 50 percent (SR_{50}) from the nontreated for the resistant (R) and susceptible (S)-biotypes measured 4100 and 221 g ai ha^{-1}, respectively. The SR_{50} for sethoxydim measured 615 and 143 g ai ha^{-1} for the R and S-biotype, respectively. The R-biotype was cross-resistant to clethodim, fenoxaprop, and fluazifop. The R-biotype was equally susceptible to the S-biotype from foramsulfuron, glyphosate, MSMA, and topramezone, averaging 62 to 75 percent biomass reductions. In laboratory experiments, the two biotypes had similar foliar absorption of ^{14}C-diclofop-methyl. Both biotypes metabolized ^{14}C-diclofop-methyl to diclofop acid and a polar conjugate, but the R-biotype averaged ~2-times more degradation than the S-biotype. Gene sequencing revealed an Asp_{2078} to Gly substitution in the R-biotype that has previously conferred resistance to ACCase inhibitors. A second mutation was identified in the R-biotype that yielded a Thr_{1805} to Ser substitution that has not previously conferred ACCase-resistance in other species.

ALLEGHENY BLACKBERRY CONTROL IN REDUCED-MANAGEMENT FINE FESCUE TURF. J.R. Brewer* and S.D. Askew, Virginia Tech, Blacksburg, VA (299)

ABSTRACT

Golf course superintendents must constrain budgets in response to reduced revenue and increased costs of chemicals, fertilizer, fuel, and other inputs. Low-maintenance turf has been increasingly adopted in out-of-play areas on the golf course to combat these budget restraints. Reduced inputs, such as biannual mowing, has led to unique weed problems in low-maintenance turf areas. One of the most prominent broadleaf weeds to encroach is blackberry (*Rubus* spp.). Blackberry is tolerant to common 3-way herbicides used on golf courses for broadleaf weed control, but metsulfuron, fluroxypyr, triclopyr, and picloram have all been proven effective on blackberry species in pasture and native areas. Our study was initiated on June 17, 2015 at two sites including: one Alleghany blackberry (*Rubus allegheniensis* Porter) (ABB) control site at the Virginia Tech golf course and a hard-fine fescue (*Festuca longifolia Thuill. 'Aurora Gold'*) tolerance site at the Glade Road Research Facility. Treatments were applied as follows: Applications were made with a CO_2 powered boom sprayer calibrated to deliver 280 L ha^{-1}, and the trial was formatted as a randomized complete block. Treatments included: triclopyr at 2.34 L ha^{-1}, fluroxypyr at 2.92 L ha^{-1}, metsulfuron at 35.0 and 70.0 g ha^{-1}, triclopyr at 2.34 L ha^{-1} + metsulfuron at 35.0 g ha^{-1}, fluroxypyr at 2.92 L ha^{-1} + metsulfuron at 35.0 g ha^{-1}, penoxulam + sulfentrazone + dicamba + 2,4-D at 7.01 L ha^{-1}, triclopyr + sulfrentrazone + 2,4-D + dicamba at 4.68 L ha^{-1}, carfentrazone + 2,4-D + mecoprop + dicamba at 5.80 L ha^{-1}, quinclorac + sulfentrazone 1.12 kg ha^{-1}, 2,4-D + fluroxypyr + dicamba at 4.38 L ha^{-1}, picloram + 2,4-D at 9.34 L ha^{-1}, picloram + 2,4-D at 9.34 L ha^{-1} + metsulfuron at 35.0 g ha^{-1}, and an untreated check for comparison. All metsulfuron containing treatments received 0.25 percent v/v of NIS. For the ABB control site, cover and control were rated every 2 weeks until 8 weeks after treatment (WAT), and then rated monthly until leaf drop. The fine fescue tolerance trial was rated for cover and injury every 2 weeks until turf had completely recovered from the treatments.

Initial ABB cover ranged from 20 to 90 percent. At 4 WAIT, all treatments containing triclopyr and fluroxypyr controlled ABB 98 percent except for 2,4-D + fluroxypyr + dicamba which controlled ABB 88 percent. Triclopyr + sulfentrazone + 2,4-D + dicamba, carfentrazone + 2,4-D + mecoprop + dicamba, and picloram + 2,4-D + metsulfuron controlled ABB between 70 and 78 percent 4 WAIT. Treatments containing triclopyr, fluroxopyr, or metsulfuron controlled ABB better than other treatments at 8 WAIT. Triclopyr + metsulfuron, triclopyr alone, and fluroxypyr + metsulfuron controlled ABB at least 98 percent by season's end. Three and four-way herbicides like Speedzone had some of the fastest activity at 1 WAIT, but by 8 WAIT regrowth of ABB had reduce control to below 60 percent. No treatment caused unacceptable injury to fine fescue. Assessments will be made next summer to determine long-term ABB control by these treatments.

TRICLOPYR INFLUENCES MESOTRIONE EFFICACY, ABSORPTION, AND TRANSLOCATION IN SMOOTH CRABGRASS. J. Yu* and P. McCullough, University of Georgia, Griffin, GA (300)

ABSTRACT

Triclopyr reduces foliar bleaching from mesotrione and may enhance efficacy for controlling multi-tiller smooth crabgrass. The objectives of this research were to evaluate the influence of triclopyr on the efficacy, absorption, and translocation of mesotrione in multi-tiller smooth crabgrass. In field experiments, tank-mixing triclopyr (560 or 1120 g ae ha^{-1}) with mesotrione at 140 g ai ha^{-1} applied sequentially or at 280 kg ha^{-1} applied singly provided excellent control (>90 percent) of multi-tiller smooth crabgrass in tall fescue. These treatments were more effective than mesotrione alone and fenoxaprop at 195 g ai ha^{-1} that averaged 66 percent and 81 percent control after 6 wk, respectively. Mesotrione alone at 280 g ha^{-1} bleached crabgrass 53 percent at 2 weeks after initial treatment (WAIT), and was 14 percent greater than the 140 g ha^{-1} treatment. Sequential treatments of the low mesotrione rate bleached crabgrass shoots 16 to 22 percent from 3 to 5 WAIT, but bleaching was <5 percent from the high rate after 4 wk. Triclopyr tank-mixtures reduced crabgrass bleaching on all dates to <5 percent. Tall fescue injury was not detected at any evaluation date. In laboratory experiments, smooth crabgrass reached peak foliar absorption of ^{14}C-mesotrione at 168 and 24 hours after treatment (HAT, 15 percent and 23 percent) when mesotrione was applied with triclopyr at 0 and 1120 g ha^{-1}, respectively. Triclopyr reduced translocation of radioactivity by 12 percent at 72 and 168 HAT, compared to ^{14}C-mesotrione alone. Enhanced foliar uptake of mesotrione from triclopyr tank-mixtures may improve control of multi-tiller smooth crabgrass, as compared to mesotrione alone. Reduced foliar bleaching from triclopyr is associated with less translocation of mesotrione or derivatives by smooth crabgrass.

CONTROL OF FALSE GREEN KYLLINGA IN COOL-SEASON TURF. C. Mansue* and J.A. Murphy, Rutgers University, New Brunswick, NJ (301)

ABSTRACT

False green kyllinga (*Kyllinga gracillima)* has become a troublesome weed of landscape and sports turfs in southern New Jersey. Herbicide options to control this weed in cool-season turfs are more limited compared to the options that have adequate safety on warm-season turfs. Greenhouse and field trials were conducted to assess the efficacy of herbicides on false green kyllinga at rates that have safety on cool-season turfs. The greenhouse trial evaluated six herbicides and five herbicides combinations for efficacy on two clones of false green kyllinga collected from two counties in NJ. The field study was initiated on 11 June 2015 on a soccer field in Ocean County, NJ having well established mats of false green kyllinga (87 percent cover in control plots on 29 Sep. 2015). Repeat applications of halosulfuron, mesotrione and triclopyr, and a single application of imazosulfuron provided 98 percent or greater control of false green kyllinga in the greenhouse trial. Repeat applications of sulfentrazone and single application of combination herbicides containing sulfentrazone provided no more than

63 percent control of false green kyllinga in the greenhourse trial. Halosulfuron and imazosulfuron were the only herbicides that provided commercialy acceptable control of false green kyllinga in the field trial. One application of imazosulfuron reduced cover of false green kyllinga to 4 percent cover by 29 September. Two applications of halosulfuron reduced false green kyllinga cover to 13 percent by 29 September; whereas, one application of halosulfuron reduced cover of false green kyllinga to 40 percent. Neither mesotrione, triclopyr nor the combination of these herbicides were effective at suppressing false green kyllinga in the field trial. Sulfentrazone applied at the greatest label rate for cool-season turf and in three of the four herbicide combinations produced some suppression of false green kyllinga (47 to 55 percent cover on 29 September). Results indicate that imazosulfuron has promise for the control false green kyllinga in cool-season turf; further work is needed to evaluate the consistency of control.

FALSE GREEN KYLLINGA CONTROL IN COOL-SEASON TURF. S.D. Askew*, J.R. Brewer, and S.S. Rana, Virginia Tech, Blacksburg, VA (302

ABSTRACT

False-green kyllinga (Kyllinga gracillima, FGK) infestations appear to be expanding in the mid-Atlantic states. In cool-season turfgrass, the most common herbicides available for weedy sedge control are sulfentrazone and halosulfuron. Both of these herbicides have been reported to control FGK inconsistently. Imazosulfuron is a new herbicide recently marketed for sedge control in cool-season turfgrass. More information is needed to develop strategies to deal with the rising incidence of FGK in cool-season turf. Studies were conducted at the Pete Dye River Course of Virginia Tech on a mixed tall fescue/KY bluegrass rough infested with 60 to 90 percent FGK. Treatments were initially applied on June 17, 2015 and repeat treatments, where applicable, were applied 6 weeks later. Herbicides were delivered in 280 L/ha spray solution using 11004 Turbo Teejet Induction nozzles to uniformly spray 1m x 2m plots. Halosulfuron (Sedgehammer Herbicide) was applied once or twice at 70 g/ha, sulfentrazone (Dismiss Turf Herbicide) was applied once at 280 g/ha, sulfentrazone + quinclorac + 2,4-D + dicamba (Q4 Plus Herbicide) was applied once at 1500 g/ha, imazosulfuron (Celero Herbicide) + MCPA + fluroxypyr + dicamba (Change Up Herbicide) was applied once at 420 and 2000 g/ha, respectively, and imazosulfuron (Celero Herbicide) + fenoxaprop + fluroxypyr + dicamba (Last Call Herbicide) was applied once at 420 and 368 g/ha, respectively. Nonionic surfactant (Induce) was mixed at 0.25 percent v/v with all halosulfuron- and imazosulfuron-containing treatments. Sulfentrazone controlled FGK 80 to 81 percent between 5 and 10 days after initial treatment (DAIT) but FGK recovered and was controlled only 53 percent and 28 percent at 21 and 40 DAIT, respectively. Unlike sulfentrazone, which exhibited maximum FGK control at 10 DAIT, halosulfuron and imazosulfuron exhibited peak FGK control at 21 and 63 DAIT, respectively. Halosulfuron controlled FGK 83 percent at 21 DAIT but not more than 31 percent at 63 DAIT and beyond. When applied twice at a 6-week interval, halosulfuron controlled FGK 99 percent at 63 DAIT and 80 percent at 91 DAIT. Imazosulfuron + fenoxaprop + fluoxypyr + dicamba controlled FGK 94 and 93 percent at 63 and 91 DAIT, respectively. The other imazosulfuron-containing treatment controlled FGK 94 percent at 63 DAIT and 77 percent at 91 DAIT. These data suggest that imazosulfuron could be an important new product for FGK control and that sulfentrazone and halosulfuron will likely require repeat treatments for effective full-season control.

MAPPING SPRING DEAD SPOT FOR PRECISION TURF MANAGEMENT. D.S. McCall*, D. Sullivan, and C.D. Shelton, Virginia Tech, Blacksburg, VA (303)

ABSTRACT

Prevalence of bermudagrass in the upper mid-Atlantic and Northeast is increasing because of improved cold-hardiness in newer varieties. Successful management of bermudagrass in marginal regions of adaptation is limited by the development and persistence of spring dead spot (SDS). Patches frequently appear in the same locations from year to year, but are difficult to control. Fall applications of tebuconazole are common on golf courses because it is economical and typically effective at suppressing SDS over time. The use of tebuconazole is not an option on athletic field turf because of label restrictions. Alternative active ingredients may have greater efficacy but are used less often because of cost. Maps to monitor SDS epidemics may be useful for precision turf management. Fungicide applications based on geographic severity can allow turf mangers to limit total treatable acreage and reduce expenditures. Additionally, such maps may help turf managers identify underlying problems that contribute to SDS, therefore increasing the probability of successful management with cultural practices. The objective of this research is generate a reliable estimation of SDS using equipment-mounted sensors that can be used in precision turf management. Geo-referenced SDS patches were collected from ten fairway locations in Virginia (VA) and North Carolina (NC) in 2014 and 2015 to create a base map of known SDS locations. Reflectance data were continuously collected from the same fairways using a cart-mounted Holland Scientific ACS470 in the red (670nm) red-edge (730nm) and near-infrared NIR (760nm) and transformed using the ratio vegetation index (RVI: NIR/Red). These data were paired with known SDS coordinates, with a random subset of ground truth samples retained for accuracy assessments. RVI averages of known locations were used to estimate SDS across all sampled and un-sampled areas. Accuracy of predicting SDS with RVI ranged from 36-75 percent in NC and 72-94 percent in VA across all fairways over two years. Observed SDS patches appeared smaller in NC than VA, though patch diameter was not measured. A possible explanation for lower accuracy in NC could be attributed to outside interference of green bermudagrass in the sensor field of view (45cm x 8cm), resulting in an RVI average that more closely resembles healthy turf. This may by reduced using higher spatial resolution, which is possible using aerial imagery collected from an unmanned aerial vehicle (UAV). One test fairway of SDS estimation in VA was divided into 66 5m x 1m grids. Approximately 30 percent of the area had no disease, with an additional 10-20 percent having low disease incidence (≤2 patches per grid). This research demonstrates that equipment-mounted reflectance mapping provides a rapid and accurate estimation of SDS epidemics, based on a small sampling of known patches. Future research will explore the practicality and reliability of fungicide reductions based on these maps.

VEGETABLE AND HERB CROPS (II)

MANAGEMENT OF HERBICIDE RESISTANT WEEDS IN NEW YORK PROCESSING CARROT FIELDS.

J.R. Kikkert* and R. Bellinder, Cornell Cooperative Extension, Canandaigua, NY (304)

ABSTRACT

New York State produces approximately 1,200 ha of processing carrots for the canned and frozen food market, with a value of roughly $9 million annually. Weeds cost growers hundreds of thousands of dollars because of improper carrot root formation, reduced yields and harvest interference. The backbone of weed control has been multiple post-emergence applications of linuron, starting with a 0.5X rate (280.2 g a.i./ha) when the carrots have at least one fully-developed true leaf and 1X rate (560.4 g a.i./ha) to carrots having 3 or more leaves. Pigweed (*Amaranthus powellii* S. Wats.) and common ragweed (*Ambrosia artemisiifolia* L.) have become particularly difficult to control in the 600 ha Potter muck region of Yates County, NY, where roughly 160 ha of carrots are grown annually. In 2012 and 2013, seeds were collected from mature pigweed plants that escaped management efforts. Seeds were germinated in a greenhouse, and 7-10 cm seedlings, were found to be highly resistant when challenged with different rates of linuron. In an effort to find alternative herbicides for growers to use, field trials were conducted in commercial carrot fields on the Potter muck (Palms muck; approx. 25 percent organic matter) during the past three years. Pre-emergence applications of S-metolachlor (1098.4 and 1490.7 g a.i./ha), ethofumesate (1681.3 g a.i./ha), pendimethalin (1681.3 g a.i./ha), pyroxasulfone (112.1 g a.i./ha), acetochlor (739.8 g a.i./ha), and prometryn (2241.7 g a.i./ha) were tested. Ethofumesate and pendimethalin did not control pigweed whereas S-metolachlor was mediocre. The best pigweed control was achieved with acetochlor or pyroxasulfone applied preemergence, followed with a single post-emergence application of 420.3 g a.i./ha linuron at the carrot 2-4 leaf stage. Less than 10 percent crop stunting was initially observed, but there was no difference in harvest yield. Prometryn was recently labeled in New York and was only tested in the 2015 trials. Concurrently, post-emergence applications of acifluorfen (16.8 and 33.6 g a.i./ha), fomesafen (5.6 and 11.2 g a.i./ha), oxyfluorfen (61.6 and 119.9 g a.i./ha), fluthiacet-methyl (2.8 g a.i./ha) and prometryn (1120.9 g a.i./ha) were tested. All post-emergence treatments in the muck trial resulted in moderate to severe necrosis to 2-leaf carrots, however this was generally outgrown and did not significantly impact yields. Data from the trials will be used to obtain new product registrations for use on carrots in New York.

PARTHENOCARPIC CUCUMBERS FOR COMMERCIAL PICKLE PRODUCTION IN THE MID-ATLANTIC. G.C. Johnson*, University of Delaware, Georgtown, DE (305)

ABSTRACT

Parthenocarpic cucumbers set fruit without pollination. Cultivars have been developed for commercial use and are common in protected culture. However, the use of parthenocarpic cucumbers for the pickle industry in the Mid-Atlantic has been limited due to the lack of adapted cultivars with desired processing traits. Trials were conducted at the University of Delaware from 2010-2015 on advanced parthoncarpic breeding materials from 4 different commercial sources. Small plot research station trials from 2012-2014 and large plot grower trials in 2014 showed that 2 parthenocarpic cultivars yielded equivalent to or better than standard gynoecious varieties with acceptable processing traits for once over mechanical harvest. These have been released as the commercial varieties "Gershwin" and "Bowie". Performance of "NQ5007" and "NQ5543" were more variable and excessive vine growth was an issue. Subsequent trials with Nitrogen fertilizer rates showed that these cultivars and sister breeding lines require 40-60 percent less nitrogen than other pickle cultivars. Parthenocarpic cultivars are grown at populations that are 50-60 percent lower than current gynoecious hybrids due to their ability to set more fruit per plant and require no bees for pollination. Wider adoption of parthenocarpic cultivars is expected. Current recommendations for production in the Mid-Atlantic region will be discussed.

EFFECT OF NANOPARTICLES OF CUO ON FUSARIUM WILT OF WATERMELON. W.H. Elmer*, The Connecticut Agricultural Experiment Station, New Haven, CT (306)

ABSTRACT

Nanoparticles (NP) of Cu, Mn, Si, Ti, and Zn oxides were applied foliarly and as a root drench (100 or 1000 mg/l) to watermelons to determine their effect on growth and Fusarium wilt. Half of transplants were inoculated with 50 ml of 50,000 spores/ml of *Fusarium oxysporum* f. sp. *nivuem* (FON). Disease progress curve was computed and plants tops were weighed. Roots and leaves were acid-digested for elemental composition. Nanoparticles of CuO and TiO applied as a root drench at 100 ppm reduced the AUDPC, but only plants treated with NP of CuO had more biomass. Only Cu was increased in the leaves of non-inoculated plants treated foliarly with both rates of NP CuO.Â Another similar experiment was conducted where NP of Cu, Mn, Si, Ti, and Zn oxides were compared to the bulked oxide equivalents. When applied foliarly, only NP of CuO reduced the AUDPC.Â In these studies, no treatment affected fresh weights.Â NP of CuO were compared to the bulked CuO equivalent, Kocide 2000, a Cu-octanoate, or no treatment in field plots. Seedlings were exposed to the treatments twice in the greenhouse before field transplanting. Seedlings were inoculated with spore suspension. Field plots exhibited very low levels of disease and there was no difference among treatments.Â However, yield did differ among treatments. Plots receiving NP of CuO had more fruit yield than the controls.Â Digests of the fruit flesh did not show any differences among the treatments for Ca, Cu, Fe, Mn, P, S, Si or Zn. Np of CuO may have value in disease management.

DOES PYROXASULFONE HERBICIDE HAVE A FIT IN SNAP BEAN PRODUCTION?

D. Lingenfelter*, M. VanGessel, B. Scott, and Q. Johnson, Penn State, University Park, PA (307)

ABSTRACT

Field studies were conducted in 2015 in Pennsylvania (Rock Springs) and Delaware (Georgetown) to evaluate herbicide programs containing pyroxasulfone for annual weed control in snap bean (*Phaseolus vulgaris*) and evaluate snap bean safety to pyroxasulfone. Studies were arranged in a randomized complete block design with three replications. The first study evaluated PRE and POST herbicide programs. Snap bean (var. 'Cassidy' in PA and 'Caprice' in DE) was planted on May 19 in PA and June 24 in DE. Most PRE treatments contained pyroxasulfone (0.08 lb ai/A or 0.133 lb) or s-metolachlor (1.59 lb). Treatments also consisted of combinations with one or more of the following either PRE or POST: halosulfuron (0.0314 lb); fomesafen (0.25 or 0.313 lb); bentazon (0.75 lb); or imazamox (0.0313 lb). POST treatments were applied about 3-4 WAP and necessary adjuvants were included in the spray mixtures. Visual weed control and crop phytotoxicity ratings were taken periodically throughout the growing period. Snap bean yield data were also collected. The second study evaluated snap bean variety response to a PRE application of pyroxasulfone (0.08, 0.133, and 0.266 lb); pyroxasulfone plus fluthiacet premix (0.134 lb); s-metolachlor (1.59 lb) plus halosulfuron (0.0314 lb); and two different formulations of acetochlor (1.14 lb). Three snap bean varieties were planted: 'Caprice' 'Cassidy', and 'Wyatt' in PA and 'Caprice', 'Envy', and 'Wyatt' in DE. (At the DE location, the study was irrigated immediately after herbicide application.) Visual crop phytotoxicity ratings were taken periodically throughout the growing period and yield data were also collected.

In the weed control study, both locations contained large crabgrass (*Digitaria sanguinalis*), common ragweed (*Ambrosia artemisiifolia*) and common lambsquarters (*Chenopodium album*). Control with pyroxasulfone PRE varied depending on location. In PA, control of all three species ranged from 65 to 86 percent control while in DE, control was >95 percent except for the lowest rate which provided <85 percent control of the broadleaves. In most cases, control was improved with the addition of a POST herbicide application. In the snap bean variety response study, differences were noted between varieties. In PA, Caprice was the most sensitive to the treatments, with Cassidy and Wyatt responded similarly. The s-metolachlor plus halosulfuron treatment caused the least injury (14 percent) to Caprice while pyroxasulfone at 0.266 lb caused the greatest (30 percent). The same trend held true with the other varieties but injury ranged from 7 percent to 23 percent. Yield did not necessarily correlate well with the visual injury ratings. Across the varieties yields ranged from 1.2 to 4.4 ton/A. In DE, crop injury (percent stunting) in general was greatest in Envy followed by Caprice and Wyatt. The pyroxasulfone treatments caused 40 to 91 percent injury to Caprice, Wyatt, and Envy. Other treatments caused no more than 18 percent injury. Due to severe injury from many of the treatments, no yield data was collected. In summary, there was a variety response to the treatments. However, as expected soil type and moisture (i.e., irrigation/rainfall) seem to have a major influence on crop injury and weed control. Pyroxasulfone may have a fit on heavier soils but on coarse soils, potential snap bean injury would be a concern.

VEGETABLE WEED CONTROL WITH PYROXASULFONE. B.H. Zandstra*, C.J. Phillippo, and M.A. Goll, Michigan State University, East Lansing, MI (308)

ABSTRACT

Pyroxasulfone is an isoxazoline herbicide classified in WSSA Group 15 and HRAC Group K3, very long-chain fatty acid inhibitors. It is labeled for preemergence weed control in field crops for control of annual grasses and broadleaves. Its low use rate and broad weed control spectrum make it an ideal herbicide for management of weeds resistant to glyphosate, ALS inhibitors, and triazines. Experiments were conducted over 5 years to determine pyroxasulfone safety on vegetable crops. Various rates of pyroxasulfone were applied to vegetable crops on mineral and muck soils to determine crop tolerance and weed control.

Pyroxasulfone was applied preemergence to carrot (*Daucus carota* L.) on muck soil at 0.15, 0.3, and 0.9 kg/ha. At 0.15 kg/ha, there was initial carrot stunting which was visible into July, but no reduction in yield. When pyroxasulfone was applied at 0.3 kg/ha, carrot yield was reduced about 40 percent. At 0.9 kg/ha, carrots did not germinate. Pyroxasulfone applied preemergence to seeded onion (*Allium cepa* L.) on muck soil at 0.15 kg/ha generally was safe with no yield reduction. At 0.3 kg/ha, onion yield was reduced 18 percent. When pyroxasulfone was applied to onion at the two leaf stage at 0.15 or 0.3 kg/ha after pendimethalin preemergence, onion yield was not reduced. Pyroxasulfone applied at 0.15 kg/ha to onion on mineral soil caused serious stunting and 40 percent yield reduction. When applied at the onion two leaf stage, onion yields were high due to improved weed control. Pyroxasulfone applied preemergence to seeded green onion on sandy loam soil at 0.037 kg/ha did not reduce yield. Seeded chive (*Allium schoenoprasum* L.) yield was reduced at this rate. In established chives, pyroxasulfone at 0.15 and 0.3 kg/ha gave good weed control and no yield reduction. Pyroxasulfone at 0.15, 0.22, or 0.3 kg/ka was applied to celery (*Apium graveolens* L.) before or after transplanting on muck soil. Celery was very tolerant of pyroxasulfone at all rates and timings and yield was not reduced.

Pyroxasulfone was applied pretransplant and posttransplant to broccoli (*Brassica oleracea* L. var. italica) and cabbage (*B. oleracea* L. var. capitata) at 0.15 and 0.3 kg/ha. Pyroxasulfone reduced broccoli and cabbage vigor and yield. Buttercup squash (*Cucurbita maxima* Duch.) and pumpkin (*Cucurbita pepo* L.) were tolerant of 0.112 kg/ha pyroxasulfone. Butternut squash (*Cucurbita moschata* Duch. ex Poir.) and cucumber (*Cucumis sativus* L.) were very sensitive to pyroxasulfone. Pyroxasulfone at 0.112 kg/ha caused yield reduction in bell, banana, and jalapeno pepper (*Capsicum annuum* L.) Tomato (*Solanum lycopersicum*) yield was reduced 65 percent by pyroxasulfone at 0.112 kg/ha.

Pyroxasulfone gave good control of annual grasses and common ragweed (*Ambrosia artemisiifolia* L), common groundsel (*Senecio vulgaris* L.), common purslane (*Portulaca oleracea* L.), eastern black nightshade (*Solanum ptycanthum* Dun.), hairy nightshade (*Solanum sarrachoides* Sendtner), and redroot pigweed (*Amaranthus retroflexus* L.). It was weak against common lambsquarters (*Chenopodium album* L.), ladysthumb (*Polygonum persicaria* L.), and yellow nutsedge (*Cyperus esculentus* L.).

EVALUATION OF HERBICIDE PROGRAMS IN FRESH MARKET SWEET CORN. D.E. Telenko*, Cornell University, East Aurora, NY (309)

ABSTRACT

Sweet corn herbicide programs were evaluated in an on-farm Extension demonstration site in Batavia, NY in 2015. Six herbicide treatments including two new herbicides that will be available in New York for 2016 were assessed. The field site was previously a hay crop for six years with a mixture of weed species included lambsquarters (*Chenopodium album* L), shepherd's purse (*Capsella bursa-pastoris*), redroot pigweed (*Amaranthus retroflexus*), green foxtail (*Setaria viridis*), common mallow (*Malva neglecta*), common ragweed (*Ambrosia artemisdiifolia*), velvetleaf (*Avutilon theophrasti*), barnyardgrass (*Echinochloa crus-galli*), orchardgrass (*Dactylis glomerata*), and quackgrass (*Elymus repens*).

Sweet corn was planted on May 15. Pre-emergence treatments of the premixture of *S*-metolachlor (1.605 kg ai ha^{-1}) + atrazine (1.798 kg ai ha^{-1}) + mesotrione (0.202 kg ai ha^{-1}) + bicyclopyrone (0.050 kg ai ha^{-1})(Acuron); and pendimethalin (1.596 kg ai ha^{-1}) + dimethenamid-P (0.945 kg ai ha^{-1}) + atrazine (1.12 kg ai ha^{-1}) were applied on May 16. Post emergence treatments of glyphosate (0.882 kg ai ha-1) and the premixture of nicosulfuron (0.034 kg ai ha^{-1}) + mesotrione (0.088 lb ai ha^{-1}) (Revulin Q) were applied on June 5. All herbicide programs performed similarly at the initial weed ratings on June 15, controlling 72 to 94 percent of grass species and 93 to 98 percent of broad leaf weed species as compared to untreated control. By August 12 the Revulin Q mixture, pendimethalin+ dimethenamid-P + atrazine, and Acuron programs continued to provide greater than 72 percent grass control and over 96 percent broadleaf weed control. All herbicide programs increase ear size, ear number, and yield (lb/A) in over the untreated control. There were no differences in weed management programs on yield in the SV9827 variety, but a yield loss was noted in the Incredible sweet corn variety where glyphosate drift injured plants. This study found that the new products Revulin Q and Acuron are as effective as standard programs and can bolster weed resistance management programs in fresh market sweet corn.

AUTHORITY MTZ DF: A NEW PRE-TRANSPLANT HERBICIDE FOR TOMATOES.
J. Reed*, D. Doohan, B.H. Zandstra, S.C. Weller, and M. Orzolek, FMC, Philadelphia, PA
(310)

ABSTRACT

Authority MTZ DF is a 45 percent dry flowable formulation containing 0.18 lbs sulfentazone and 0.27 lbs metribuzin. Authority MTZ DF is labeled for use as a fall application and spring pre-plant and pre-emergence application ahead of corn and soybeans. Recently Authority MTZ DF has been labeled for use pre-transplant ahead of tomatoes. Authority MTZ DF may be applied safely to provide control of such weed species as herbicide resistant and susceptible palmer pigweed, other pigweed species, common lambsquarters, velvetleaf, Pennsylvania smartweed, black and Eastern black nightshade, specific grasses and yellow nutsedge. Authority MTZ DF use rates range from 8 oz product, 0.225 lbs ai/A (0.252 kg/HA) to 12 oz product, 0.338 lbs ai/A (0.378 kg/HA) on most soils according soil texture. Authority MTZ DF is an excellent PPO pre-mix herbicide alternative to control glyphosate and ALS herbicide resistant weeds for tomato producers.

Author Index

Author	Abstract Number
Gosselin, Andre'	58
Gould, Juli	41
Gover, Art	171, 223, 227, 230
Grande, John	28
Grant, Jessica I.	105
Gray, Stewart	94
Greig, Kelly	295
Gribskov, Michael	52
Groden, Eleanor	110
Grosman, Donald M.	170, 201
Growe, Anthony M.	287
Gugino, Beth K.	62
Guo, Li	61
Hager, Aaron	187
Hall, Cordelia A.	126
Hamby, Kelly A.	268
Hamilton, Abbe V.	84
Hamilton, George C.	42, 43
Hao, Jianjun	93
Haring, Steven C.	51
Harlow, Christopher D.	161
Harpster, Tracey L.	291
Hartzler, Bob	187
Hashemi, Masoud	32, 33, 249
Hines, Thomas	191
Hoban, Jacqueline N.	104
Hoepting, Christine A.	126, 270
Hogan, Conor T.	209
Holdcraft, Robert J.	101
Homa, Kathryn	266
Hooks, Cerruti	108, 268
Hough-Goldstein, Judith	14
Hoyle, Steve	228
Hu, Shuijin	70
Huebner, Cynthia	252
Hunn, Jonathan M.	222
Hunter, Mike	141
Hunter, Mitch	179
Infante-Casella, Michelle	164
Ingber, David A.	74
Inman, Matthew D.	213
Irwin, Caleigh	295
J. Fornazier, Mauricio	128
Jahanzad, Emad	33
Jemison, John	280
Jennings, David E.	260
Jennings, Katherine M.	188
Jeranyama, Peter	236
Jernigan, Ashley B.	67

Author	Abstract Number
Jhala, Amit J.	142
Jiang, He	93
Jimenez-Gasco, Maria	87, 96
Johnson, Gordon C.	305
Johnson, Quintin	283, 307
Johnson, Steven B.	93, 271
Joline, Meghan C.	129
Jones, Sharon	115
Jones, Tappey T.	209
Jordan, David L.	70, 18, 213
Jordan, Nick	187
Jung, Geunhwa	95
Kahl, Hanna	109
Kalmowitz, Kathie	207
Kao-Kniffin, Jenny T.	242
Karayilanli, Elif	281, 282
Karsten, Heather	90
Keach, James E.	203
Keene, Clair	56
Keese, Renee J.	207
Keilsohn, William A.	174
Kenaley, Shawn C.	172, 195
Kennedy, Ashley C.	77
Kersbergen, Richard	280
Ketterings, Quirine	30, 31, 81, 141, 154, 157, 246, 247, 248
Khatchikian, Camilo E.	211
Kikkert, Julie R.	304
Kilcer, Tom	81
Kleczewski, Nathan M.	59, 144, 149
Klick, Suzanne	118
Klimaszewska, Krystyna	217
Knight, Alexandra M.	70
Knowlton, Katharine	250
Koppenhofer, Albrecht M.	72
Kraemer, Benjamin	158
Kubinec, Diane M.	281
Kuhar, Thomas P.	40, 103
Kula, Robert R.	110
Kulesza, Stephanie	250
Kunkel, Brian	118
Kunkel, Daniel	165, 166
Lagace', Marie	217
Lake, Ellen C.	14
Lamb, Elizabeth M.	162, 192
LaMondia, James A.	198, 192
Lamp, William	105
Lawrence, Joe	141
Leach, Ashley B.	35
Leskey, Tracy C.	76, 115